D0041042

Planet of Microbes

Planet of Microbes

The Perils and Potential of
Earth's Essential Life Forms

TED ANTON

The University of Chicago Press
Chicago and London

The University of Chicago Press, Chicago 60637
The University of Chicago Press, Ltd., London
© 2017 by Ted Anton
All rights reserved. No part of this book may be used or reproduced in any manner whatsoever without written permission, except in the case of brief quotations in critical articles and reviews. For more information, contact the University of Chicago Press, 1427 E. 60th St., Chicago, IL 60637.
Published 2017
Printed in the United States of America

27 26 25 24 23 22 21 20 19 18 17   1 2 3 4 5

ISBN-13: 978-0-226-35394-4 (cloth)
ISBN-13: 978-0-226-35413-2 (e-book)
DOI: 10.7208/chicago/9780226354132.001.0001

Library of Congress Cataloging-in-Publication Data
Names: Anton, Ted, author.
Title: Planet of microbes : the perils and potential of earth's essential life forms / Ted Anton.
Description: Chicago ; London : The University of Chicago Press, 2017. | Includes bibliographical references and index.
Identifiers: LCCN 2017016254 | ISBN 9780226353944 (cloth : alk. paper) | ISBN 9780226354132 (e-book)
Subjects: LCSH: Microorganisms — Popular works.
Classification: LCC QR56 .A58 2018 | DDC 579 — dc23
LC record available at https://lccn.loc.gov/2017016254

⊗ This paper meets the requirements of ANSI/NISO Z39.48-1992 (Permanence of Paper)

Contents

Introduction

Forty-seven-year-old Nora Noffke scanned the Australian desert at the end of a blistering day. With her Old Dominion University student Dan Christian, she was seeking evidence of ancient life. It was their second day in the national park, and some of the equipment was already covered with dust. They had gotten lost once. The leaves of acacia trees and grass stirred over rocks some three and a half billion years old. In the morning the air had smelled fresh and clean, but now, at sunset, the stench of a cow carcass rose from a nearby dry streambed.

It was July 2011, and they had taken four flights over three days from Norfolk, Virginia, to Sydney, and then to Perth. Before being allowed into the Pilbara National Park, she had to present her project to officials of the Australian Geological Survey. They flew to a coastal town and rented a four-wheel-drive Jeep to drive toward the dusty town of Marble Bar, famous for its horse race in July. Seventy-five hours after leaving Virginia, in the middle of the night, they reached the desert. Exhausted, they pushed ahead on a dirt rut that dwindled to nothing. They could see their mountain site in the distance but had no idea how to get there. It was so dark they finally pitched their tent at a homestead's crumbled foundation, too tired to go another mile.

Noffke studied the intricate, beautiful, multicolored microbial mats that live in the Earth's tidal pools and salt lakes. Some protected preserves surrounding the Red Sea and some Chesapeake Bay, and some appeared in ancient sediments much earlier in Earth's history than many believed complex life had existed. Other microbial biofilms helped treat wastewater. Some consumed oil spills, protected our lungs and gut, infected hospitals, and formed the plaque on our teeth. An expert in microbe mats and the sediments they altered, Noffke was one of the few researchers who believed their remains could be found in rocks in the Australian desert formed during the Earth's earliest years.

For two days Noffke and Christian had hiked, hauling water tanks, food, graphs and field journals mile after mile along the rocky scree of Saddleback Ridge. Their eyes adjusted to the peculiar red landscape populated by fire ants, snakes, and a few wild cattle. It was so hot in the afternoon they covered the tent and vehicle with a tarp and huddled beneath it. But at night, as they played cards around the campfire, brilliant stars exploded over them. Two hundred million miles away a NASA rover was soon to look for the same structures on Mars, on a landscape similar to this, though far colder.

A transplant from Germany, Noffke loved field trips with her students. Her father, an air traffic controller trained by the US Air Force, taught her that people must be reminded of life's wonders sitting right in front of them. On drives from their suburban Stuttgart home, he pointed out to her and her brother the glittering Jurassic fossils of ammonite and tiny animals in the rolling farmland. Later she studied under a supportive Harvard mentor who pioneered research into similar fossils all over the world. In her career she helped push back the dates of microbial sedimentary structures, from one to two to nearly three billion years ago, from the Middle East to South Africa and Australia. Those discoveries taught her that living microbes had inundated the ocean shallows of the earliest Earth. She had first come to Australia's Pilbara in 2008 with another geologist. On their last day she had glimpsed the mat structures—but there was no time left. She vowed to come back.

As the sun set, Dan Christian pulled out a flask of warm water. A wedge-tailed eagle swooped over the ridge in the distance. The red sun slipped to the horizon. Shadows lengthened.

Something in the boulders' shadows caught her eye. She grabbed Dan and pointed. A foot-long series of parallel ridges lay in the stones. More wavy lines appeared, astonishing and beautiful. She knew those patterns! The fragments looked as if they had grown yesterday. "Dan, look, there they are! Tons of them!" she cried. "They're rolled up even."

They raced toward the hill and skittered to their knees, digging and brushing in the dry earth, shouting and laughing in the fading light.

We are living through an unprecedented time of discovery about the origins of life. Huge strides have been made in our understanding of the steps by which life may have formed on Earth, and life-friendly worlds outside our solar system may be glimpsed in the near future. The heroes of this study are microbes, which dominated the Earth through most of its existence and created the oxygen we breathe and the biological processes, like respiration and metabolism, on which our lives depend.

The amazing thing is that relatives of the same ancient organisms that live in swamps and the acid waste of mines, in the hot vents at the ocean's bottom and the high-radiation environments of the stratosphere, also live in our own intestinal tracts and those of farm animals. They may prove of great practical use in the remediation of radiation contamination, petroleum spills, and other toxic wastes. They convert waste into fertilizer and energy and may hold the keys to confronting the increasing rates of obesity, asthma, autoimmune diseases, and even some of our most mysterious emotional illnesses.

Most of us know something about the probiotics on the shelves of any health-food store or the microbes of the human gut. Their news feeds would be hard to miss. What we may not know is that their chemistry offers clues to the processes that gave rise to life. The anaerobic conditions of our intestines resemble some of the conditions of mining wastewater and fertilizer runoff, and some of the conditions of ancient Earth or even Mars. The fascinating research into organisms that thrive in such environments, from the hot pools of Yellowstone Park to the tops of Antarctic mountains, the acid lakes of Australia and California, backyard compost heaps, and our own bodies, hold clues to solutions of some our most intractable problems. Few animals or plants could exist without their microbes. A human body has some thirty to forty trillion. Without them, we would die.

Microbes are the world's best chemists. They can take almost any mineral and derive energy or nutrients from it. They make our wine and cheese. But for years we knew little about their ubiquity and diversity, and we considered them mainly as enemies in our fight against infectious disease. Now we know they are critical to our and the planet's health. This book describes the next step in the story of a revolution in understanding: Microbes are the hidden underpinning of the global ecosystem. They live almost everywhere and can metabolize just about anything. To understand them is to offer a way out of some of the crises we have created for our world.

When rigs and tankers pour millions of gallons of oil into the ocean or pipes into soil, microbes do most of the cleanup. They created the oil, from decayed plant material, in the first place. When patients lie near death from recurrent *Clostridium difficile* infections, microbial transplants can save them. When you cannot sleep or suffer from allergies or depression, it could be that your gut microbiome is partly at fault. These minute organisms offer clues to confronting some of our fastest-growing maladies.

This is the story of the struggle to unlock and apply those clues, from 1920s experiments showing that plant and animal cells are the result of an ancient union of two microbes, to the discovery of a third form of life in hot springs and hydrothermal vents, to the critical insights into the human microbiome's role in our physical and emotional health. Our natural microbes outnumber our own cells by one and a third to one or, at least, match them. Their seven million genes far surpass the small number of our 25,000 that we use. Some of these microbes play key roles in our metabolism and mental health. Some could help us confront a warming Earth's weather, and many already offer organic ways of cleaning the environment. They offer us clues to energy sources and hints of a more sustainable future. A new microbe industry might harness these remarkable organisms, if only we could better manage their abilities.

The race to understand the microbiome takes us on journeys with people making discoveries about the Earth's early history, from volcanic thermal pools to miles below the Earth's surface, from microbreweries to Mars, from ocean bottoms to the labs where synthetic cells are being made and new medicines tested. For all our genetic and computing wizardry, we know very little about the microbial world, having catalogued perhaps one percent of the tiny organisms that house tens of millions of mystery genes.

I first wrote about microbes years ago in another book and felt then that they deserved a book of their own. Now they rank as one of the biggest stories in science and medicine, with announcements happening almost daily and one or two books appearing each year. The idea that exposure to germs helps children develop immunities was a favorite of my father-in-law, a doctor raised in India. Now we have the tools to understand how such immunities might work. As I hiked along the islands of Croatia and the cliffs of California, I felt inspired with excitement for the future. From brain disorders to obesity to antibiotic resistance to global warming and the energy crisis, our former enemies, the microbes, are now becoming allies in an all-out effort to seek better health, sustainability, and a deeper appreciation of the true diversity of life.

This book follows a three-part revolution. The first part treats the discoveries in life's origin and potential glimpses of it elsewhere in the universe. The second analyzes and applies the effects microbes may have on our physical and emotional health. The third focuses on the fundamental role microbes play on our planet, providing renewable energy and nutrients, remediating waste, and shaping our climate and environment. These related quests are transforming our understanding of the world and our place in it.

Along the way, researchers are changing the manner in which discoveries happen. New, remote-operated experimental tools, coupled with information science, have created a flood of microbial discoveries, some profound and some spurious, as discoveries push back the existence of microorganisms on Earth to four billion years ago and beyond. Some discoveries are shared so quickly they defy old barriers of analysis and review. This book delves deeply into the personalities and methods of new researchers to understand how their discoveries are changing the way science is done. Scientists are gathering big data in new ways, in clusters of centers or islands of information, almost outstripping our ability to understand their meaning, seeking patterns that enable a new visualization of the tree of life.

Few recent fields have riled up more critics and churned out more controversial figures, ranging from the University of Massachusetts researcher Lynn Margulis, who proposed a holistic understanding of the Earth's relation to its organisms, to researchers pushing back the origin of life to the planet's beginning, to the numerous biomedical leaders making claims and racing to raise money for new drugs to replace those that no longer work. Their science is forging new connections among disparate disciplines, such as medicine, geology, and chemistry, that are changing our vision of the world.

Several books have covered the new findings about the roles

played by human and animal microbiomes, including Ed Yong's *I Contain Multitudes: The Microbes Within Us and a Grander View of Life*. This book applies those findings to some of the deepest questions and crises confronting us. The human, animal, and plant worlds are inextricable from the planet's geology, climate, and environment. That microbial insight may help crack the code of life itself and reshape the future of our planet and ourselves. Where did microbial life appear, and how? What are the proper roles of our body microbes, and how might they address the increase in rates of obesity, anxiety disorders, and autoimmune disease we now confront? How might this hidden world provide new energy sources, or methods of cleaning our spills and waste? How are people trying to make money on these discoveries and what, exactly, are they trying to do?

The world's tiniest and most successful organisms were here long before us, and they will still be here long after we are gone. This is the story of the race to understand and harness their power. We live in their world.

PART ONE

Out of the Air
Searching for Life's Origin

1

Lightning in the Lab

The University of Chicago campus was freezing in December 1952 when Stanley Miller, a thin twenty-two-year-old California graduate student, walked into the basement lab. He found his unwieldy three-foot-high twin glass globe contraption still sputtering electricity from its Tesla coil. The water was simmering, and he turned down the flame to look.

The water was cloudy, and the collection globe was covered with brown gunk. Miller's heart raced as he pulled open the hatch to sample the tarry residue.

Miller was a Jewish younger brother from Oakland, California, an Eagle Scout who loved the outdoors and chemistry and had little patience for small problems. Known as a chemistry prodigy in his undergraduate days at University of California, Berkeley, he was anxiously following the war news from Eastern Europe—World War II had sepa-rated his grandparents in his native Latvia. His father, a lawyer, was appointed by family friend and future Supreme Court Justice Earl Warren to be Oakland's assistant district attorney. Then his father died suddenly in 1946, threatening Miller's dream of pursuing a doc-toral program. He needed money to follow his older brother into graduate school.

Miller had applied to the country's top biochemistry programs and waited, unable to sleep, hoping. Only one school, the University of Chicago, offered him financial support. He would have to teach.

These were the earliest days of molecular biology and the origin-of-life search. On the one hand came important papers, like those of the physicist Erwin Schrödinger's lectures collected in the book *What Is Life?*, which inspired Francis Crick and James Watson. Schrödinger explored the fact that life is the only entity that seemed to defy the law of entropy and prophetically pushed researchers to unravel the molecular basis of heredity.

On the other hand, compelling fantasy fiction, like that of H. G. Wells, promoted the conviction that life must exist elsewhere and may well be smarter or better than we were. But few really imagined that serious work into the topic could be pursued, though a Soviet naturalist, Aleksandr Oparin, and an English biologist, J. B. S. Hal-dane, had each independently written about the chemicals that would be needed for the creation of life from chaos. Sputnik would soon pour the finances of the U.S. government into pure science. At the dawn of big science, sketchy but exciting speculations received full U.S. and Soviet government support.

Miller arrived at the imposing southside Chicago campus in September 1951, a smallish, brilliant, brash, and insecure graduate student loudly eschewing the slow work of experiment as "time-consuming, messy and not as important . . . as theoretical work," he later wrote. Instead he tried theoretical physics with the controver-sial Edward Teller, who was studying the early universe, but it was difficult research that did not involve his favorite topic, chemistry. Miller floundered. After a year the government called Teller to work on the hydrogen bomb in California, and Miller had a lucrative offer to go with him.

It was cold in Chicago, and Miller longed for his native state, where he could help make weapons for the U.S. Army at high pay with benefits. One day he attended a campus lecture by the professor and pastor's son Harold Urey. The monthly lectures were high-pressure affairs that encouraged and tore down some of the best minds in the lecture hall surrounded by such Nobel Prize–winners as Enrico Fermi. Urey was a quiet, self-deprecating westerner with his own Nobel. "It is possible to create an experiment," he said at the end of his talk, "to recreate the conditions of early Earth and see what lightning, in the form of electrical discharges, might produce." Miller sat upright in his chair.

Miller was galvanized but uncertain, and it took him months to approach Urey. Miller said would try the origin-of-life experiment. Urey turned him down. Miller pressed him. He really thought the experiment might work.

Finally, reluctantly, Harold Urey permitted the Californian to try an origin-of-life experiment. He gave him one year.

A beautiful world

In the 1920s Oparin and Haldane had suggested that the chemicals of the early Earth might, if zapped with an energy source, create dynamic disequilibria, which would be a precursor to life. The physicist Erwin Schrödinger explored in his lectures the many ways in which life defied entropy, the tendency of systems to wear down over time. In this universe, nothing gained energy. Except life.

The 1950s marked a paranoid time of Communist witch hunts, atomic obsession, and the Korean War. Exceptions to the gloom were the nascent fields of space science, science fiction, and molecular biology, where all nations could meet in a dream of human better-ment. In Cambridge, England, Rosalind Franklin was X-raying DNA and James Watson and Frances Crick were deciphering its structure. Japanese movies portrayed a world united against mutant mon-sters. Science-fiction writers such as Robert Heinlein, Ray Bradbury, Ursula K. Le Guin, and Isaac Asimov kindled dreams of strange life in the universe. As a child in the 1960s and '70s I spent nights devour-ing their novels.

In Chicago, Harold Urey was a fifty-nine-year-old Indiana minis-ter's son who grew up in Montana, who first saw an automobile at the age of seventeen, and who taught science in a mining camp in Para-dise Valley, memorialized in the Jimmy Buffett song "Cheeseburger in Paradise." Urey's discovery of the element deuterium, also called heavy hydrogen, had swept him into the new physics world of radio-activity. Isotopes got him to Europe in the 1920s to meet Heisenberg and Einstein and to New York in the 1930s to study radioactive dating at Columbia. His discovery of deuterium, a key element of the atomic bomb, swept him into the secret government war effort, a high-stakes, complicated, and top-secret effort that wore out the pacifist Nazi-opponent, who told President Truman not to use the bomb.

Once the war was over and he moved to the University of Chi-cago, Urey analyzed planet climates and defended the Rosenbergs be-fore the House Un-American Activities Committee. He studied an-cient atmospheres by the relative differences in oxygen, trying to understand the early days of the solar system. Deuterium had given humankind one gift: a reliable way to date the rocks of the distant past. That was when he mentioned his idea for an origin-of-life ex-periment in a lecture.

"I already have a Nobel"

In the basement of the university building, Urey's graduate student Miller created three glass contraptions, mimicking the long-necked glass tubules employed by Louis Pasteur, to electrify the gases he thought existed on early Earth. He fashioned three types of the same experiment: the volcanic version, the lightning version, and a straight version. "Water is boiled in the flask," he wrote, "mixes with the gasses, circulates past the electrodes, condenses and empties back into the flask."

It was a struggle, but Miller's mentor, Urey, understood what it was to struggle. When Urey's first dissertation topic fell through at the University of California, he went to work with the great Niels Bohr in Copenhagen on the spectroscopic study of molecules. But the secret, high-stress atomic bomb effort of the 1930s and '40s had worn Urey out, and his proposal for the bomb's triggering device was turned down. Urey almost skipped his own Nobel Prize ceremony to be at the birth of his daughter.

In 1952, at the University of Chicago, Miller had the experimental tools made and walked in to see his experiment. Within two days the liquid turned a pale yellow, and in a week it was marred by "cloudiness and turbidity." The cloudiness came from organic material mixing with the silica from the glass. In a laborious process that would identify its composition, Miller dipped a paper in the tarry residue and dipped it first in alcohol, then in phenol, and finally in another chemical. The spot turned purple, which meant glycine, an amino acid, was present. It was a small amount but still, a building block of life he had made in two days of an experiment they both thought would never work. Miller raced to call his older brother Donald.

Glycine is a building block of proteins and of pharmaceuticals. What Miller did not know at first was that as his dipped mixture turned pink, the mixture was building a small portion, some 2 percent, of amino acids, the building blocks of living cells, as well as other biomolecules such as the hydrocarbon bitumen. He had glimpsed a potential vision of the origin of life.

Miller repeated the experiment while Urey was out of town on a lecture tour, this time sparking the mixture for a week. The inside of the flask became coated with an oily scum, and the water turned brownish. Now the glycine spot was far brighter, and other amino acids showed up as well.

Once Urey returned, the result became clear to them, and it was explosive: amino acids appeared in conditions that resembled those of early Earth. When Miller repeated the experiment a generation later, he created thirty-three amino acids, including half of the twenty found in proteins. The acids appeared in surprisingly consistent doses, and some were keys to life. Having learned the power of publicity from his work on the atomic bomb, Urey pushed his graduate student to write it up fast. Miller produced a draft was surprised to get it back from his professor, with a long list of comments, in only a week. By February 1953 it was done. Urey removed his name and sent it in himself to *Science*. "I already have a Nobel," he said to Miller.

Weeks passed. Urey wrote the editors to question the delay. Another month passed. Urey sent a telegram to the *Science* editor he knew; *Science* did not appreciate it. In a panic, Miller tried another journal, the *Journal of Biological Chemistry*, which accepted it at once. Then *Science* came back on March 25. A reviewer apologized to Urey for the delay.

The article appeared two months later, a few weeks after the announcement of the double helix for DNA. Amino acids appeared in conditions Miller and Urey thought resembled those of early Earth. The two appeared in the *New York Times* and became famous overnight when the tabloids erroneously reported that they had invented life. The British journal *Nature* had published the double helix structure of DNA, and this was the American *Science*'s answer. Radio and newspaper reports telephoned the men for quotes. But the hardest part for Miller was his assignment to deliver the monthly lecture to a skeptical group of famous University of Chicago faculty that he had heard Urey give. Now he was the main actor. Enrico Fermi stood and fired back that he thought it all seemed very unlikely. Urey backed his student. "If God did not do it this way," he said, turning to Fermi, "then He missed a good bet."

Miller need not have feared the school faculty. Soon he would make the cover of *Time* and hear his name mentioned as a prospect for a Nobel Prize. With his simple experiment a new field was born: astrobiology. But on the other side of the world a strange thinker was coming at life from the opposite direction, wandering the foothills near his beloved Russian lakes. His work had everything to do with microbes.

Foothills of lichen

When the scientist awoke he walked down the rotting wood blocks to the lake. He would toss his shirt onto the broken lawn chair on the dock and sink into the water. The fish would scatter, his toes would grasp onto a rock here or there and push off, and he would swim. Konstantin Merezhkovsky would surface to the lonely calls of loons.

At the turn of the century, microbes were still the enemies of humankind. Microbe hunters such as Louis Pasteur and Robert Koch had defeated cholera, tuberculosis, and anthrax, all while racing one another, drumming up funds from national governments and wealthy patrons, and laying the foundation of modern medicine, often at risk to their health and reputations. By the 1940s it seemed like an end to disease was in sight.

There was only one problem. In so doing we were disturbing ancient relationships that had protected us for hundreds of thousands of years. Only a handful of the microbes in our bodies are killers. Many are not only beneficial but in fact critical to us and to the biosphere. Without them, we would not be here.

Few researchers were interested in the beneficial microbes save for a group of idealistic Russian thinkers studying some of the benefits tiny living beings—from algae to parasites, intestinal bugs to those on our skin—provide for us and other animals. The naturalists studying them called themselves the Russian School. Of them, one obscure, troubled individual wandered the lakes, foothills, and steppes of Kazan province, studying the hard-shelled creatures called diatoms that made pond scum and fouled well water. Shaped like elongated diamonds, diatoms often lived in symbiosis with fish and other marine creatures. They floated in such huge packs in the sea that their skeletons formed the White Cliffs of Dover. Konstantin Merezhkovsky studied their role in the biosphere while penning children's fantasy books geared toward his favorite audience, young girls.

Gradually Merezhkovsky shifted his attention from diatoms to the strange combination of microbe and fungus that made up common forest mosses and lichen. These commensal organisms seemed to suggest a big idea: symbiosis, or living together in biochemical cooperation, made for a major factor of life in Russian forests. The union of independent organisms is a life strategy that drove evolutionary advancement. Cooperation, not competition, was the model for complex plant and animal survival.

Merezhkovsky understood that lichen, or moss, was not the plant it seemed, but rather a cooperative community of algae and fungi. You could see them in a microscope. He collected hundreds of lichen species from all over Russia and Austria, and around the Mediterranean, studying them their medicinal uses as best he could. In his spare time he wrote fantasy stories of ritual initiation and secret societies. He published his scientific papers on symbiosis in journals and then in a book. Multicellular life grew out of the ancient unions of bacteria that performed essential roles to aid each other.

In his own country Merezhkovsky was criticized, while few abroad noticed. Yet there was nothing new about the idea of symbiosis. Darwin anticipated it in 1868 when he wrote: "Each living creature must be looked at as a microcosm—a little universe, formed of a host of self-propagating organisms." All around us is evidence of symbiosis. Ancient cyanobacteria that the made the pungent blue-green algae of swamps and ponds, Merezhkovsky proposed, were the basis for chloroplasts, the photosynthesizing engines of plant cells that cleaned our air, created our oxygen, and helped to generate our soil and forests and prairies.

Without pond scum, in other words, life and the environment as we know them would not exist. Merezhkovsky suggested yet a bigger, more audacious idea: the nucleus of plant and animal cells, the control center whose rules all complex cells followed, was itself an independent bacterium incorporated in an ancient union of microbes. So were other vital plant and animal cell organelles. Indeed, the modern cell was, as Darwin intuited, a microcosm of diverse, formerly independent, intricately cooperating parts.

To most people, bacteria—some round, some rod-shaped, some skinny, some fat—caused disease, spoiled food, and signaled decay. Fungi were synonymous with foot and body secretions we hated to consider. Found in rocks and crevices around the world, they went unnoticed because of their mysterious lifestyles in soil and dead things, and as symbionts on animals, plants, bacteria, or other fungi.

Yes, some microorganisms were hugely beneficial. Yeast was a common one-celled fungus that made bread rise and beers, wines, and liquors ferment. Bacteria performed essential roles in recycling organic matter and cycling nutrients that allowed plants to thrive. Still, the dominant model of life's development was of the majestic tree of visible life sketched by Darwin and filled out further by others. There was little room on it for secret sharers.

The strange Merezhkovsky gathered the ideas of the Russian School together in a series of closely argued papers and then quietly left Russia. He moved around the world under a pseudonym, the author of racist and anti-Semitic tracts. Someone was following him. Convicted of raping several girls in Russia, the married aristocrat escaped to the United States, where he worked for two years at Stanford under the name William Adler and tried to seduce a farmer's daughter. He fled to Austria, then to Germany, then back to Russia until finally, penniless, he committed ritual suicide by imbibing poison, as in a scene from his children's novel.

It was not surprising his biological ideas found only a small audience. Outside of the civil war–torn Soviet nation, few read Russian and fewer still paid attention to the ideas of a psychopath. How odd that such a flawed individual, building on the ideas of others, could offer one beautiful insight.

One person with much the same insight was from a remote mountain town in Colorado.

"A new principle in organic evolution"

One person who independently came up with the same idea as Merezhkovsky was the avid mountaineer and eccentric medical school professor Ivan Wallin. A child of Swedish immigrants raised on an Ohio farm, Wallin attended Illinois's Augustana College and the University of Iowa as an undergraduate, earning his doctorate in anatomy at New York University. In Boulder the intimidating Wallin taught at the University of Colorado Medical School, where in his spare time he thought about the energy units called mitochondria in the human cell. Why did mitochondria have their own genes, he wondered.

On campus Wallin was a tall, striking figure, frequently rushing to do his experiments on the cells of leftover beef from the butcher, never lecturing in his classes but rather demonstrating by dissecting human cadavers and interrogating students as they watched. If a student answered incorrectly, the professor might punch him in the chest. On weekends these same students helped him build a mountain cabin twenty miles north of Boulder in North St. Vrain Canyon, nicknamed Club Wallin, where their professor hunted and fished and hosted an annual Christmas glogg party featuring poker, carol singing accompanied by Wallin on the piano, and such Scandinavian delicacies as pickled herring salad, lutefisk (dried cod soaked in water and lye), and glogg, a high-octane drink of mulled wine and aquavit.

Although Wallin lectured little, he wrote up his experiments on trying to culture mitochondria independently in a rapid and bold succession of eight scholarly articles. He announced his theory that the complex cell's mitochondria were once free-standing bacteria in a book he wrote in 1927, *Symbionticism and the Origin of Species*. "A principle that is so revolutionary as Symbionticism," he announced boldly in his preface, "not only introduces a new principle in organic evolution, but also inserts a new conception in heredity, development and cell-structure." All animal and plant cells originated in a union of microorganisms.

True, he wrote, "a large number of biologists have been skeptical . . . and many have been opposed" to his conclusion. He never could cultivate mitochondria independently in the lab he created in a shed behind the medical school classrooms. That would be impossible because mitochondria draw some of their genetic capability from a mammalian cell nucleus. The mitochondria he claimed to have cultured outside the cell were shown to be the results of contamination. Wallin practiced medicine, taught his classes, and tried to do pure research, all while administering a medical school. Dubbed "Mitochondria Man" by his opponents, the disgraced Wallin eventually gave up his microbe research to concentrate on teaching and school administration.

The idea of symbiosis as a driver of evolution flew in the face of the Darwinist school of linear, random genetic variation. Few researchers gave it much credence. The scientific tide went to the mastery of the gene, under the control of each individual's autonomous DNA. The power of the Nobel-winning Miller-Urey experiment and the beauty of the Watson-Crick DNA double helix of the same year swept through biology and popular culture. Step by step, molecular cryptographers unraveled the gene's amazing qualities of control. On comets and in the dust clouds of deep space, NASA probes detected some of the same protochemicals used by Miller and Urey. The implications were so forceful that they blinded some of the research that followed. "Stanley Miller inspired a new generation of younger scientists," the Carnegie Institute geologist Robert Hazen told me, "but spent his later career tearing down anyone who challenged him." When carbon was discovered in meteorites, it seemed as though we were on the verge of finally understanding life's origin. Life competed; the world went to the strongest. There was no incentive for a selfish gene, or ancient bacterium, to cooperate.

Wallin's and Merezhkovsky's ideas would come back a half century later and change the face of science. It is now biological dogma that the cells of animals and plants were born in the union of ancient microbes, and that such complex symbioses might hold potential clues to our health, longevity and even the planet's future. But no one would have thought it at the time.

Then a vexing and brilliant woman became interested.

The Instigator

From her home in a working-class bungalow on Chicago's South Halsted Street near the stockyards, Lynn Alexander loved going to the public library with her father. Her family, descendants of Zionists, was scraping by on her father's law practice until he died suddenly at the age of fifty-one. Lynn still played on the Chicago rooftops and back fire escapes and learned the piano, but she felt her mother struggling to make it. At the University of Chicago Laboratory High School, the daughter found an escape in reading great biology papers. She had to discuss a question: "What is the living material basis of heredity?" That question inspired her to pursue her interest in biology, as it did her fellow Lab School student James Watson.

Finding few boys, however, she left the Lab School for public high school. She immediately felt she had made a terrible mistake. At home, in an act of rebellion motivated by her new stepfather's constant fighting with her mother, at the age of fourteen she applied to the University of Chicago. Argumentative and self-assured, she got in.

At the university she ran into a tall, dark-haired, good-looking eighteen-year-old physics major wearing dark socks and shorts and bragging that he would be the first scientist to find extraterrestrial life. Pretty, with English-schoolgirl features and kindness for anyone who needed help as well as impatience with anyone she considered wrong, Margulis challenged him. It was Carl Sagan. Confounding him with her intelligence, she graduated from the University of Chicago at age nineteen and, against everyone's advice, married Sagan two weeks later. At age twenty-two, the pregnant Margulis earned a master's degree in biology from the University of Wisconsin–Madison. She wrote her thesis on the microbe *Euglena*. She went for her doctorate at Berkeley with an infant in tow. When her stepfather, a gambler, squandered $500 in a night of poker, a fortune to her hard-pressed family, Margulis bought a Mayan statue of the same value, which she willed back to Mexico on her death.

Harvard hired her brash husband, and she followed with her toddler and new infant. Not content to be a faculty wife, however, Margulis applied for and won an instructorship at Boston University. She staked her ground watching strange microbes in the world's hidden, odorous swamps and effluents. She studied the amazing complexity and genetic ingenuity of their weird abilities to metabolize hydrogen or iron and manganese and live in the most inhospitable environments. She coined the term "microcosmos" for this hidden ancient life, so fundamental to visible life, reminding readers who ignored microbes that this "planetary patina . . . has continued for more than three billion years." Margulis wrote up her research on such obscure topics as spirochetes, bearers of syphilis, and viruses in the digestive system of termites. Early microbial life evolved by sharing genes in communities that became the model for and precursor to the organelles of the modern cell and its abilities to metabolize, photosynthesize, and reproduce. Symbiosis, she argued, was the major force of life's innovation.

In 1966 she penned a paper pulling together the work of Wallin, Merezhkovsky, and others to make the point that photosynthesis, the Earth's great oxygenator, spread when an early microbe, the blue-green organism in tidal pools and algal blooms, was incorporated into another cell to become the life-giving chloroplast of green plants' cells. Mitochondria, the energy factories of both animal and plant cells, were once freestanding bacteria. Cell division of plant and animal cells was a choreography of ancient free microbes. Rejected by sixteen journals, that paper, "On the Origin of Mitosing Cells," came out in 1967, the year she divorced Carl Sagan and kept custody of their two sons. The compelling, well-argued paper on symbiosis, the first to link Wallin's and Merezhkovsky's ideas, was at first mostly ridiculed or ignored.

Undaunted, Margulis sold symbiosis in the 1970s and '80s, drawing on her brilliance, her passion, and the revolutionary understanding that most of the capabilities of plants and animals derived from bacteria. They make our oxygen and soil, recycle everything that dies, and create most of the processes, such as respiration and metabolism, on which our lives depend. As she moved on to a tenure-track position at Brandeis University, she became a scourge of hierarchical science and a role model for rebellious thinkers. Most important, she offered a vision of the profound power of microbial life in the world.

"Like a pointillist painting"

The medieval tree of life featured two branches, plants and animals. Then Anton van Leeuwenhoek turned his handmade microscopes to rainwater and butcher's meat and a whole new world of tiny "animalcules" opened to the human gaze. "I have seen," van Leeuwenhoek wrote in 1696, "an unseen world. My work . . . was not pursued in order to gain the praise I now enjoy, but chiefly from a craving after knowledge." The Dutch lens maker wrote long letters to London's Royal Society and to Admirers around the world. He entertained kings in his humble dry-goods store but never revealed his best techniques and tools, even though Europe's nobility clamored to see the "incredible number of little animals . . . which move very prettily . . . tumble about and sideways, this way and that!"

From that came the nineteenth-century discoveries that such scourges as anthrax, cholera, gonorrhea, tuberculosis, the plague, and syphilis were the work of microbes similar to those that had fascinated van Leeuwenhoek. The rivalry between the Frenchman Louis Pasteur and the German Robert Koch and the public health crises they confronted produced the breakthroughs identifying the tiny culprits behind devastating fatal illnesses, such as the plague which killed up to one-third of Europe in its great medieval sweep. In the early twentieth century the Rockefeller Foundation journalist Paul de Kruif, writing about those triumphs, popularized the term "microbe" for the germ killers that were seen as the main enemy of health. Little distinction was made, observed Margulis, between deadly and benign germs, viruses, or bacteria. When Alexander Fleming discovered that a compound in bread mold could destroy pathogens, modern antibiotic medicine was born. But in all those triumphal quests, researchers miscalculated. They thought microbes were mostly enemies of humankind, an idea advanced in treatises "by authoritative specialists, usually large men of commanding presence," Margulis later wrote. Only a few thinkers looked to the benefits they provide for plants and animals, and these scientists mostly toiled in obscurity.

More than a hundred million of them might live in a spoonful of seawater. They lived three billion years before us and will continue long after we are gone. In a thriving culture, if you start with one, in twenty minutes you may have two; in another twenty minutes, four. Three times an hour such a microbial population could double, so that in twenty-four hours they might overwhelm a small square of tidal beach or a human body. In America today, treating infectious disease costs $120 billion, but the cost of not doing so is even greater. That's why Koch, Pasteur, Fleming, Jonas Salk, and Albert Sabin worked so hard to find ways to kill them. Shaped like squiggles, rods, circles, polyhedrons, spaceships, blobs, and trapezoids, some emit the swamp stench my seventh-grade science teacher called "the smell of life!"

Most biologists studied plants or animals, while microbiologists focused mostly on cell proteins or nucleic acids. Unless researchers were studying disease, few wanted to pay attention to microbes.

However, the idea of symbiosis galvanized some scientists, who understood that in the vast scheme of life on Earth, many microbes are beneficial, some very much so. They make much of our oxygen, especially if you add in the plant chloroplasts that descended from a microbe, regulate atmospheric gases that shape our global climate, and form not only the cilia and flagella of vital pond and ocean plankton at the bottom of the food chain base but also the flutes of tree leaves and the stripes of Dutch tulips. They help to make sea animals and fireflies luminesce, mammal brains grow, and the human immune system work. With about twenty-four ways of breathing, they were far more adaptable than animals, which have only one. Beautiful and strange, microbes keep our planet in its delicate, life-sustaining balance, even while we do everything we can to disrupt that balance.

Margulis understood what it was to fight to survive, living with her three younger sisters in one room while her parents argued in the next. Margulis became a supporter of young people and felt an affinity for the beauty of the homely and forgotten. Her interests outside of science ranged from Emily Dickinson's poetry, to Colombian and Mayan art, to her beloved dogs, one named Velasquez, after the painter. She worked on a kibbutz in Israel at age fourteen and conducted field research on Mexican village at age twenty.

Margulis researched, published, and raised her children, marrying again and divorcing the crystallographer Thomas Margulis. "Motherhood, marriage and science do not mix," she later observed. At age thirty-two she expanded her ideas on symbiosis in a book titled *Origin of Eukaryotic Cells*, summoning all the known data—in a time before DNA sequencing—to give evidence for symbiosis as a driver of evolutionary change. The title made sly reference to Darwin's *Origin of Species* and forged an erudite technical style meant to intimidate opponents. It also announced her role as public provocateur.

When her mother, Leone, died of a stroke in 1982, Lynn Margulis worked harder. In 1981 she had published *Symbiosis in Cell Evolution*, marshaling new evidence from the discovery that cell mitochondria retain their own DNA, which enhanced the supposition that were likely once freestanding. As she explained it in a *Discover* magazine interview with her Amherst neighbor, the science writer Dick Teresi:

All visible organisms are products of symbiogenesis, without exception. The bacteria are the unit. The way I think about the whole world is that it's like a pointillist painting. You get far away and it looks like Seurat's famous painting of people in the park. Look closely: The points are living bodies—different distributions of bacteria. . . . Symbiogenesis recognizes that every visible life form is a combination or community of bacteria.

Her vision of life as a pointillist microbial painting began to catch on with a disparate group of thinkers, from fields as diverse as ecology, microbiology, marine biology, climate science, and astrobiology. But few acknowledged its significance.

Shunned by many of her field's academics, she forged a lucrative career as a popular scientist that also helped to advance her ideas. When she criticized NASA for not understanding how to look for life on other planets, it made her the first woman to receive astrobiology funding and then promoted her to direct its planetary program. Turning then to the search for life in the solar system, she finally won recognition when she was admitted to the National Academy of Sciences in 1983. She was hired by the University of Massachusetts, where she became a beloved professor and mentor to women and men in science. She continued her field research on microbe mats in swamps, bogs, and tidal pools while penning articles and books on symbiosis, sex, and evolution that offered a fresh understanding of life's development. Yet she struggled to reach an audience, at least among scientists.

"A revolutionary idea whose time has come"

In our soil live millions of different microbes, lichens, fungi, and tiny insects. Inside the insects, plants, and animals live millions of vastly different microbes that help them to breathe, digest, fight infection, and maintain well-being. What Darwin's followers could not explain, Margulis argued, was the imaginative novelty of those life forms. Neo-Darwinists, the name she gave to popular master-gene theorists such as Richard Dawkins, proposed that single random gene mutations gave us life's unbelievable diversity, including, among almost countless other forms, the eight-to-ten million different types of microbes. Dawkins and most other evolutionists countered that small, random mutations promoted gradual change, such as the beautiful gradations of the Galapagos Island finch beak as each generation adapts to varying amounts of rainfall. That view, Margulis argued, could not explain the explosion of new species, of dinosaurs and mastodons, palm trees and jellyfish. Only symbiosis could. The reason we have so much variety in life was that the microbes had been furiously gene swapping for eons, she said. The ways that milk cows digested, Pacific coral reefs grew, and New Guinea plants survive is through symbiosis with microbes. Life depended on these biochemical associations.

While Margulis sampled the muddy, colored microbial mats in Cape Cod tidal pools and volcanic vents, she also studied the spirochetes that cause syphilis and the wood-eating symbionts in the intestines of termites that enabled them to destroy houses. She loved the malodorous microorganisms that others shunned and used them to shock her audience. "Much of the evolutionist's modern terminology should be abandoned," she wrote with her son Dorion Sagan. Bacteria were so interdependent with their symbionts that they almost could not be considered as separate species. "Identity is not an object, it is a process," she wrote, meaning that such functions as metabolism and respiration preceded the fixed structures that carried them out. "More bacteria went to the moon in the Apollo missions," she pointed out, "than humans." Zoologists like Dawkins and Stephen Jay Gould missed the vast majority of life and its history. It was because of symbiosis that we are here.

Richard Dawkins called her "Attila the hen," and Gould debated her, but others resonated to her ideas. The Spanish biologist Monica Sole Rojo became a fan and confidante when she saw Margulis face down Gould at Oxford. If Margulis could do it, Sole Rojo reasoned, so could she. A porn actor, Connor Habib, was inspired by Margulis to study evolutionary and organismal biology.

Gould and Dawkins ridiculed the claim that symbiosis drove the significant leaps of natural selection. As a paleontologist who, like Margulis, funded his research by speaking and writing for high fees in the public sphere, Gould thanked Margulis for standing up for microbes but never once mentioned symbiosis in his encyclopedic *Structure of Evolutionary Theory*. Dawkins, in his popular book *The Selfish Gene*, described symbiosis "as a revolutionary idea whose time has come" yet said little more.

When the speed of genetic analysis improved, however, evidence accumulated. In the 1980s proof of symbiosis came after genetic analysis showed that the reason mitochondria had their own DNA was that they were indeed once freestanding organisms and that chloroplasts were once blue-green algae. In "one of biology's great ironies," noted author Michael Gray in *Nature*, our energy-producing mitochondria are descended from typhus-causing *Rickettsia*. By 1989 the idea had gained enough ground that Margulis organized the world's first international conference on symbiosis, funded by a Rockefeller Grant, in Bellagio, Italy. She invited new researchers and old friends and thinkers from around the globe.

The time for symbiosis had arrived, and additional evidence supported its validity. Bacterial nodules on plant roots enabled them to extract nitrogen from soil. One mouse's gut bacteria changed its behavior to make it lose its natural fear so that the host could be eaten by a cat, in which the microbe reproduced. Other microbes in the human gut helped us to produce the vitamins we needed to survive. Life operated in networks, something like the people Margulis helped.

When scientists rejected her, she brought her ideas directly to the public. Then her ex-husband introduced her to his English office mate at the Jet Propulsion lab in Pasadena, California, the inventor James Lovelock, and the significance of the relationship between microbes and the planet changed forever.

"Gaia is a tough bitch"

James Lovelock was an only child in a lower-middle-class Quaker family in rural England who worked in industrial research before getting his doctorate. He had a knack for invention and a contrarian way of thinking that overcame his self-doubt and innocence. During World War II he studied infectious disease for the Medical Research Council in London bomb shelters, where he once had to help deliver a baby. When the young mother was asked who the father was, she replied, "I'm not sure, it was just a come-and-go in the shelter." The shy Lovelock recalled losing his virginity in much the same way.

Lovelock specialized in inventing detectors. He improved the gas chromatograph, a device that could identify the components of a gas using, among other methods, the spectrum of light passing through it. He created a detector to identify the air movement from such disasters as volcanoes, nuclear tests, and industrial and radioactive breakdowns. These inventions propelled him into the emerging major science battles over the environment and into the big popular publications *Nature* and *Science*; at the same time, he maintained his practical approach to science, honed by his experience in industry. The British and U.S. governments and such companies as Shell and Rothschild hired him to exploit his ingenuity, enabling him to leave university teaching and live an independent, quiet life in the rural north English village of Bowerchalke.

His most famous invention was the electron capture detector, which he used to detail the worldwide atmospheric spread of industrial poisons such as polychlorinated biphenyls (PCBs) and chlorofluorocarbons (CFCs) on a 1972 voyage following the path of Darwin's *Beagle*. Other scientists realized that CFCs were causing a growing ozone hole in the Earth's atmosphere, leading to the first environmental explosion and successful government regulations of aerosols, and marshaled his research in a worldwide campaign. The device was among the most sensitive of chemical analytic methods ever invented. But when Lovelock realized on a subsequent voyage that the crew's spray antiperspirants were fouling his data, he politely changed his research rather than disturb the crew.

His research on atmospheric imbalance attracted the attention of NASA, and in the late 1960s it invited him to design detectors for its Mars probes. Unlike the other biologists and geologists on the team, Lovelock reasoned that the best way to search for life would be to seek an unstable atmosphere, one that was out of equilibrium owing to the metabolism and respiration of living organisms. NASA was so enthralled it made him the director of its Voyager probes until Congress canned the missions. When Lovelock and Carl Sagan were temporary NASA office mates, the English instrument designer was writing a paper on the biosphere's effects on planetary atmospheres. Microbes in the soil and the ocean and in the form of chloroplasts produced Earth's most volatile and valuable atmospheric element, oxygen, and controlled its abundance. They regulated the planet's temperature and the saltiness of its oceans. Ocean algae, Lovelock suggested, made sulfur compounds that seeded rain clouds. Earth life helped to shape the makeup of the Earth's atmosphere. In Pasadena he was struggling with the full implication of his ideas. Sagan put him in touch with his ex-wife.

In polite correspondence in which the two highly independent thinkers addressed each other as "Dr. Margulis," and "Dr. Lovelock," they discussed their shared notion that a life-Earth system regulated the planet's climate and geology. After they met at a conference, Margulis provided Lovelock's hypothesis with microbiological depth, not to mention enthusiasm, confidence, and connections. Lovelock traveled to meet Margulis, and she insisted he stay at her Amherst home, not the hotel he had booked. Then he had a heart attack. She and her second husband rushed him to the hospital, and from there the relationship was set.

Margulis's microbial expertise brought the missing biological depth to Lovelock's idea of a "geobiological cybernetic system." Lovelock quipped that Margulis was interested in a "three-billion-year-old septic tank," but she pointed out that those same foul-smelling microbes shaped most of the Earth's current life-supporting atmosphere. She hammered home the full power of his hypothesis, writing to him in a letter: "This is a new scientific paradigm, on the order of Kuhn"—Thomas Kuhn, the philosopher who coined the term "paradigm shift." Together Margulis and Lovelock proposed the heretical notion that microbial life had modified Earth's climate and environment for billions of years.

Lovelock invoked an unwieldy name: "a cybernetic system with homeostatic tendencies." Prodded by Margulis's criticism, he asked his neighbor William Golding—author of *Lord of the Flies*—for a catchier suggestion. The novelist suggested the term Gaia, for the Greek goddess of Earth. At first it seemed like a terrific title for his theory. But over the years the mystical overtones hindered its acceptance. For better or worse, however, the name has stuck: the Gaia hypothesis.

The Gaia hypothesis holds that the Earth and its organisms, mainly the microbes, formed a self-regulating system that shaped the environment to be suitable to life. Microbes created our oxygen as waste, maintaining the life-giving gas at the right percentage—21 percent of Earth's atmosphere—to sustain animals and plants. Ocean algae may help form the clouds that protect us from the sun's heat. Nitrogen fixers help plant roots obtain nitrogen. Microbes produce most every gas in our atmosphere. As a triggering part of the rising environmental awareness of the 1970s, Gaia helped inform the whole-earth movement and the fields of ecology and environmental studies.

The idea was widely misunderstood to argue that the Earth somehow was alive. Scientists lampooned it in the 1980s as a trendy fad. The biologist John Maynard Smith called it an "evil religion." Others criticized it as lacking hard data. Many simply would not "discuss it in polite circles," Margulis observed. When *Science* rejected an early paper, Lovelock replied, "The editorial board must be senile." In an expanding series of conference presentations and professional and popular articles and books, the two refined the theory in a partnership propelling each other to fight back. "Life can flourish only within a narrowly circumscribed range of physical and chemical states and since life began the Earth has kept within this range," Lovelock wrote. "This is remarkable for there have been major perturbations such as a progressive increase in solar luminosity, extensive changes in the surface and atmospheric chemical composition and the impact of many planetesimals." Microbes engineered our planet to sustain them, even as they evolved to meet environmental and geological changes as they occurred. In some two dozen papers and book chapters, Lovelock continued to develop the Gaia idea from a shaky hypothesis to a full-fledged theory that could, potentially, be proved or disproved.

"Gaia is a tough bitch," Margulis liked to say before launching into a series of pointed questions of lecturers. Speakers learned to wince when she came barreling to the microphone to question them.

Rounding the dark side of the moon in 1967, Apollo 8's astronauts snapped the famous "Earthrise" photo of a sapphire globe whiskered by white cloud and ice and blotted with brown and green continents. The image shaped pop culture and burned the idea behind Gaia—that the Earth and its life forms form a powerful interconnected whole—into the popular imagination so vividly as to become a cliché. Featured in rock songs and head-shop posters, it showed the frighteningly fragile interplay of life and Earth. "Gaia," said one of Margulis's graduate students, "is symbiosis seen from space."

One well-heeled organization attuned to Margulis's and Lovelock's ideas continued to be NASA. After it put Margulis in charge of its new planetary biology division in 1982, she chaired the National Academy Committee charting planetary projects for more than a decade. A graduate student working for her, the Duquesne University environmental microbiologist John Stolz, recalled debating a creationist and answered the growing number of phone and mail requests. "I was her aide-de-camp. It was an unbelievable experience," he recalled from his Pittsburgh office years later. "We never thought outside the box because, with Lynn, there was no box!"

Margulis told NASA it should point its high-tech sensors at our planet, where our sweet life-supporting air came because microbes ate up the carbon dioxide and produced the oxygen. Without microbes, Earth would be almost as hot as Venus. Lovelock predicted the disappointing results of the Viking landings, because it was known that Mars's atmosphere was at an equilibrium. The Lovelock strategy of detecting atmosphere disequilibrium is now a common feature of the search for life on planets outside of the solar system. The Margulis idea of turning NASA's sensors on Earth is a controversial political topic of our day.

A game-theory experiment offered perhaps the best proof of the principle of Gaia. Lovelock described in a paper a place called DaisyWorld, with two types of daisies, one dark-colored and cold-loving but heat-absorbing, the other light-colored and heat-loving but light-reflecting. On a cooler Earth, dark daisies would thrive until such point when the heat they trapped would favor the lighter-colored flowers. Light daisies would take over, eventually cooling the Earth until the dark-colored twins again thrived and the process was reversed.

In the real world, the new science of climatology gave the Gaia theory its biggest support and most compelling hard data: for years the new interdisciplinary theory had been invoked to analyze the dangers of global warming created in large measure by human activity. In a swirl of conferences and competing claims the Gaia theory became the intellectual underpinning for a social and political movement.

"A minor twentieth-century religious sect"

Margulis spent her last decade as she had the previous five, in combat with the major figures of her field. With her son Dorion she wrote more books expanding on her ideas, including *What Is Life?* (2000) and *Acquiring Genomes: A Theory of the Origin of Species* (2008), drawing on the gene revolution to argue that the reason we found it so easy to swap genes and engineer crops was that real-life microbes had been doing it for generations. She continued her passion for the tiny, ugly, and ignored microbes of swamps and landfills, seeing in them potential answers to life's biggest questions.

In Amherst she was a familiar, eccentric figure who bicycled to work and skinny-dipped in Puffer's Pond. In a famous Oxford College showdown, nicknamed the Battle of Balliol, she finally faced Dawkins. She would take her students to Sippewissett Swamp in Massachusetts in knee-length waders and fleece hoodie, pulling up the mud to show the layers of multicolored microbes, orange, blue-green, and pink (for diatoms), with the cyanobacteria and sulfate-reducing anaerobes at the bottom. The Gaelic biologist Denis Noble, author of *The Music of Life*, serenaded her in medieval Occitan on guitar. She lived next door to Emily Dickinson's house at 20 Triangle Street and read the poet closely. Margulis wrote a novel about scientists having affairs, based on her ex-husbands. In homage to J. D. Salinger, she signed her letters, "With love and squalor." At a conference, meeting the noted evolutionist Niles Eldredge, she asked him to dine at Chuck E. Cheese so she could bring a grandchild. The two thinkers scrunched into little plastic chairs amidst the life-size metal cartoon characters to share the latest science gossip.

She could be angry and sweeping and sometimes wrong. Symbiosis was not just a factor—it was the main driver of evolution, she wrote. Neo-Darwinists wore "male capitalistic, imperialistic" blinders who missed the essential act of sharing that watered the tree of life. She likened them to "a minor twentieth-century religious sect within the sprawling religious persuasion of Anglo-Saxon biology." She thought spirochetes, the sources of syphilis, gave rise to the protists' flagellum.

But her primary scientific idea galvanized a newer and more careful generation of thinkers. These included the University of Hawaii zoologist Margaret McFall-Ngai and Duquesne University's John Stolz, among many others. Independently, these two made discoveries that augmented the idea Margulis championed: that microbes are major shapers of Earth's geology and weather. Climatologists studying global warming seized on the idea that the planet and its organisms constituted a single entity. The Earth's first great toxic event, as Margulis and Lovelock often argued, was indeed the bacterial creation of oxygen, deadly to most of Earth's early microbes. Those beneficial organisms retreated to rank swamps, hot vents, and animals' and our intestines, where today they may affect our mental, physical, and emotional well-being.

Stolz felt her work was "exactly the kind of thing I wanted to do," he recalled, "to understand how living things" interact with minerals and climate. He built a prominent career on the uses of microorganisms to remediate petroleum, fracking, and other toxic wastes, as well as to generate energy from municipal trash. The leading climatologist Stephen Schneider was inspired to organize a Gaia conference in 2004. Dorion Sagan founded a global Gaia Society with a Facebook following. Microbes were found in every inch and corner of the Earth, it argued, from toxic waste sites to miles below the Earth's surface to the heights of the radiation-saturated stratosphere, and symbiosis could be tapped.

Other scientists, like Caltech's Dianne Newman, followed up on the symbiosis ideas in a lab investigating primitive metabolism and cystic fibrosis. When Margulis was admitted to the National Academy of Sciences she renewed her studies of bizarre bacteria in tidal pools, swamps, and sulfurous hot vents as the causes of syphilis and helpers of termites and cows. Newman went onto become a major figure in the study of microbes and human health but was so shaken by the attacks on Margulis that she once emailed Hawaii's Margaret McFall-Ngai to ask about her originality. McFall-Ngai replied that Margulis was a theorist "whose contribution should never be forgotten!"

At the height of her career, Margulis had four grown children and three grandchildren and was coauthoring a book, *Dazzle Gradually*. In it she argued for a completely new theory of the origin of species, a problem Darwin never quite solved. New species arose when microorganisms joined. It was not biological dogma that tiny organisms joined in an ancient union that led to modern plant and animal cells. More significant than many of her books, it argued that most gene variation did not arise from random mutation, as so many had thought, but from integrating partial or even whole genomes of free microbes. The ability to rapidly swap useful genes is what makes pathogens so quick to resist our medicines. The result of thirty years of research, it brought the main points of her host of discoveries together, leading Harvard's E. O. Wilson to describe her as "one of the most successful synthetic thinkers in modern biology." She won science awards in Russia and Europe.

She could overstate her various cases, as many pointed out and, as a theory Gaia is not widely accepted, though its underlying system of holistic interpretation reshaped policy. At home, it is safe to say, she was never fully recognized.

New vision

I am walking along the tidal pools of Crystal Cove State Park in southern California, beneath the sandstone cliffs that stretch to the Baja Peninsula, pounded as they have been for millennia, imagining Charlton Heston in the closing scene of *Planet of the Apes*. In the bay surfers await the next big wave as the Beachcomber Grill sounds the bell for martini sunset. Beneath the crumbling bungalows, I think of Margulis. The cells out in those bays use quantum mechanics to sense the Earth's magnetic field. The mitochondria of my cells were descended from freestanding bacteria, perhaps in tidal pools similar to these. Those ancient microbes created the oxygen and many of the other gases in our atmosphere. They process our climate and affect the weather. They alter minerals that reshape the Earth's surface. Many of Earth's significant features spied from space, from massive marine clouds formed around microbially induced matter to the Grand Canyon's banded red iron stripes, were affected by the actions of microbes.

Margulis and Lovelock were among the first to see those connections, and she was among the first to use the term "human microbiome" for the human microbial community, pointing out that we live not as individuals but complex entities. "When she talked about the human microbiome it fell on deaf ears! I worked on biofilms as a grad student. We called them mats," recalled Stolz. "We'd go to these meetings and there would be a handful of people. And then the message got through that biofilms are essential to human health and now you can't get into those sessions."

Margulis was an adept synthesizer of ideas who wrote directly and with polemical joy to the public that funded scientific research. Eventually Schneider chaired three conferences to discuss the pros and cons of Gaia and became a major popularizer, in print media and on TV, of global warming. "If you don't want to play the game," he once told me about public speaking, "it doesn't mean you are somehow pure. It means you've abdicated!" Brian Rosborough, the founder of EarthWatch, the world's largest and oldest nonprofit organization that pairs supporters with ecology-minded research trips, called Lynn Margulis a "life force." Numerous students and scientists, many of them women, looked to her as a mentor.

As for Lovelock, he also experienced a new level of success in his late seventies. After his long-ailing wife passed away, he fell in love with new passion and remarried. He won four major international awards and authored two popular books. But in a world focused more on competition than cooperation, his ideas on the fundamental importance of microbes to health, climate, and Earth sustainability were mostly ignored as a New Age fantasy. The feedback systems between microbes and Earth "were largely accidental," argued Dalhousie University's W. Ford Doolittle in a typical rebuttal. There was no way to legislate against cheats, suggested Dawkins, such as plants that produced no oxygen but derived all the same benefits as those that did. Genes affect an organism, not the entire planet's weather.

While Margulis and Lovelock became science celebrities in their later years, they also became a bit cranky or went completely off target in some public pronouncements. Before her death in 2009 Margulis offered that AIDS did not result from HIV and repeatedly suggested that the attacks of 9/11 were engineered by a right-wing conspiracy. Lovelock largely rejected the green movement as hopeless and saw little hope for government action in combating global warming. His idea that earth's oxygen level has remained constant at 21 percent has been debunked.

Nonetheless, Margulis's and Lovelock's ideas changed public policy and our understanding of the Earth as an interconnected system, opening a rich new interdisciplinary field of study that shaped the environmental movement, energy and public policy, astrobiology, and even medicine. The biggest questions could be answered by analyzing the smallest of creatures. Gaia emphasized the profound interaction of microbes with geology and climate and our own behavior. Numerous studies gave support to its major points, including the discovery of a Great Oxygenation Event 750 million years ago, when the oxygen produced by cyanobacteria transformed Earth's atmosphere and made complex life possible on the planet. DDT's devastating effects on bird populations, coupled with carbon emissions' damage to our air and the discovery of the growing ozone hole, helped lead to the creation of the U.S. Environmental Protection Agency. Gaia's system of interpretation succeeded, even if its science was challenged.

Long before probiotics became popular with the general public

Margulis advocated eating kefir, the thousand-year-old Middle East-ern cultured-milk drink. She rejected science's metaphors of gene control, of acquisition and contamination, favoring instead a kind of microcosmic interaction. "My work," she often said in typically polemical style, "crossed the boundaries . . . of thirty disciplines." She used language that left her colleagues unsure of where the indi-vidual ended and the community began. A fan of the science major Emily Dickinson, at the end Margulis was no longer writing science but rather polemical personal narrative. Multiple voices of future sci-ence leaders, even those who disagreed with her, took up the call, as younger thinkers sought to find hard data to understand the power of mutualism.

To be sure, many disagreed. The University of Washington pale-ontologist Peter Ward argued forcefully that life produced as many chemicals to harm other forms of life as it did to benefit them, and that for every commensal community there is a vicious war zone of organisms seeking by any means to kill one another or not caring one way or the other. The conflicts produce chemicals and behaviors de-signed exactly to harm or extinguish other forms of life. Ward pointed to the same Great Oxygenation Event as one piece of evidence: the oxygen of cyanobacteria that made dinosaurs and humans possible also killed off most extant life over the millions of years during which it took place, shunting the microbes we call extreme into niches like swamps, animal intestines, deep mines, and vents. Oxygen was one of the world's great pollution crises.

There matters remained, in an argument about science metaphors and a philosophical stance that many felt lacked solid proof. Half a century later, however, the Whole Earth movement would be a world-wide phenomenon of research, articles, websites, lifestyle choices, government policy changes, corporate sponsorships, and a vital shift in scientific thought. Margulis linked "two previously unconnected facts," the Heinrich Heine University botanist Bill Martin summa-rized to me. "Chloroplasts were originally freestanding bacteria, as were mitochondria." If you could understand the microbial world, added the University of Illinois evolutionist Carl Woese, "you could understand how life came to be on the planet, how it functions on the planet." But microbes were still the enemy, and most doctors and drug manufacturers sought to defeat them.

Then two breakthroughs opened a new vision of the cell's origin.

3

In the Hot Vents with RNA

At the University of Colorado in Boulder, the chemist Tom Cech was a tall, bespectacled mentor with a knack for handling people. An avid skier and later the father of two daughters, he would sometimes head into the Colorado mountains to "ski really fast" and "let loose." In 1980 he was thinking about an experiment he was doing by himself.

As a junior high school student in Iowa City, Cech would walk over to the University of Iowa's stately white-limestone buildings and knock on the doors of professors to ask questions about things like crystal structures in minerals. His father, a physician, grilled him on the workings of the flowers and animals of the Iowa River valley. As an undergraduate at Iowa's Grinnell College, he loved studying Homer's *Odyssey* and Dante's *Inferno* as much as chemistry, and he married his organic chemistry partner. He later tried to emulate those Iowa professors in his own dealings with students.

By the early 1980s origin-of-life research had slowed considerably. For a generation after the seminal Miller-Urey experiment, origin-of-life research had focused on the prebiotic soup. Stanley Miller redid his original experiment and showed you could make almost three dozen amino acids by varying the makeup of the model atmosphere. But they could never get to phosphates and sugars, the real stuff of life. They could not come close. After the rush of early discoveries, the emotional roller coaster slid down as the years passed, and Miller's view sank to a low point. "I confess I see no solution," Miller had written of the problem of making nucleic acids or sugars, though he would later change his mind. By the early 1980s, much of the excitement about life's origin had fizzled.

Cech had little initial interest in the origin of life. He tried advanced chemistry research in his graduate work at Berkeley and chafed at its slow pace. At Colorado he was studying the material that wraps the DNA of our genes, called chromatin. He was trying to understand how chromatin affected transcription, the process by which DNA information is copied into RNA to create proteins. Cech was studying ribosomal RNA (rRNA), critical to making proteins, and had discovered a specific part of the rRNA that needed to be removed before rRNA could do its job. Studies of RNA splicing, the removal of its nonworking parts, had just begun, and Cech was having trouble with his experiment.

Unbeknownst to him a team of three well-known researchers — Leslie Orgel at the Salk Institute, Francis Crick of DNA fame, and Carl Woese at the University of Illinois — had proposed that RNA, the genetic messenger, might be the first step toward life's origin. Focusing on a specific splicing reaction that intrigued him, Cech was doing the experiment by himself, without telling anyone, in between his teaching duties. The splicing reaction was so specific and repeatable he pulled out his organic chemistry textbook to understand what was going on. It had to be the result of a contaminant protein, he thought, but Cech's team tried for a year to identify the enzyme they thought was doing the splicing. "This can't be right," Cech told his assistant, Art Zaug, after multiple failures.

Suddenly the Cech lab had a heretical idea. All the cell's chemical reactions need to be accelerated, or catalyzed, to occur fast enough for life. At the time, all known biological catalysts were a type of protein called an enzyme. Could it be that RNA could also act as a catalyst for biological reactions? Perhaps the reason he could not find a protein enzyme was that there was no enzyme—the RNA molecule itself was doing the enzyme work. It was acting like a protein. Cech and his team had discovered that some types of RNA could do the work of enzymes, catalyzers of the cell's key reactions. They dubbed these RNA molecules "ribozymes." RNA had the potential not only to carry genetic information, but also to put it to work in the cell.

This had huge implications for understanding how life began. If RNA could store genetic information *and* trigger life's chemical reactions, it indeed could be life's essential precursor. Few believed them. Luckily, though, at Yale, a Montreal-raised chemist was making the same discovery.

"Fight for what you believe"

Sidney Altman was born in 1939 in Montreal into a family of poor Russian Jewish immigrants, his mother a textile-mill worker and daughter of a rabbinical scholar, his father a grocery-store clerk. He grew up close to the same blocks as the author Saul Bellow and, later, the songwriter Leonard Cohen. Two childhood events sparked Altman's awe of science. First was the atomic bomb, shrouded in its destructive, mysterious, violent power. Second was the gift of a book he received at thirteen on the periodic table of the elements, which captivated him with the beauty and symmetry of the physical world's order. In his Notre-Dame-de-Grâce neighborhood, meanwhile, the eccentric immigrant small businessmen taught him the value of an education, street wisdom, hard work, and, most of all, "to fight for what you believe."

Altman studied physics at the Massachusetts Institute of Technology, where he took a course in molecular biology. He was enthralled both by MIT's excitingly high-stress atmosphere and by revolutions in the field. He made it to Colorado for his doctorate, where he learned that education could be fun, not stressful; the thinker George Gamow was a supporter. An avid writer, Altman penned stories for the children of his mentor, Thomas Puck. In 1962 Altman read the *Nature* paper by Brenner and Crick on the triplet code, by which the messenger RNA makes a template that is read in the cell's ribosome to make proteins. Captivated, he joined the microbiologist Matthew Meselson's team at Harvard University and then the group led by Sydney Brenner and Francis Crick at the Medical Research Council laboratory in England. There Altman helped discover the first precursor to a transfer RNA molecule.

That discovery earned him a post at Yale in 1971, where he became a professor in 1980. Altman was studying the properties of a bacterial RNA-protein hybrid and suspected that RNA alone could do the work of the protein. Like Cech, Altman discovered that RNA would splice itself and act as an enzyme to catalyze its own reactions. He too came under fire from competitors and friends alike. One biologist called him a crank.

Altman survived in part by his Montreal "innate stubbornness" but at a party confessed to his mentor that he was worried that "nobody believes this." Meselson asked if he had done all the proper controls. Yes, he replied, he had.

"Then that's what nature is telling you," Meselson said.

Prior to 1989 RNA was thought to be only a passive genetic carrier—it bundled information that told where to go and what to do to help foster protein synthesis. Both Cech and Altman discovered that when left alone, some types of RNA will splice themselves and act as enzymes. Most of the essential components required for enzyme action could be found in RNA, which meant that RNA might be a driving force behind the origin of life instead of just a bystander. They also discovered that most of the essential components required for enzyme catalysis could also be found in RNA structures. It was like answering the age-old chicken-or-egg paradox with a third contender, present in both and responsible for both. Fortunately for both men, they made the same discovery in different model systems. The Nobel committee called it "a complete surprise."

A new vision, dubbed the RNA world, was that a nucleic acid might have done the original work of microbial metabolism. The RNA world hypothesis stated that early life on Earth used RNA both to store information and to act as an enzyme to catalyze its own reactions. RNA behaved as the genetic keeper and messenger. In the modern cell many types of this versatile acid, including messenger RNA, transfer RNA, ribosomal RNA, and small nucleic RNA, as well as long-noncoding RNAs—each perform critical functions, some not well understood even now. The RNA world hypothesis suggested that this versatility made RNA it an ideal first step toward the earliest forms of microbial life, which could have depended completely on RNA to store genetic information and trigger chemical reactions, since most of the essential qualities of a protein are also found in RNA. The hypothesis cast RNA in a much more pivotal role in the creation of life than previously thought.

Today, RNA biology is a forefront of life-science research. The small RNA snippets that block gene transcription have led to an industry based on RNA interference.. With more than two billion dollars in investment, several biotech companies are racing to turn RNA into a new kind of drug. HIV, Ebola, the flu, and the common cold are RNA viruses. Altman theorizes that interference in the action of RNA could help to devise a new approach to medicine, and Cech and others are following the universe of different RNAs, in the microbiome and in human cells, that can be used to modify the expression of genes affecting human longevity. RNA had many functions, DNA only one. The discovery that RNA could splice itself offered a radical reinterpretation of life's origins, reinvigorating the quest for microbial precursors, as Harvard's Walter Gilbert wrote in his essay coining the term "RNA world": "RNA is the catalyzer and synthesizer of all life."

But there remained several problems with the RNA world. RNA was a highly unstable messenger in the elaborate genetic machinery of the modern cell. It also had many other functions, and both RNA and DNA seemed too complicated to have originated on their own out of a few amino acids floating randomly in an ancient ocean. They were information processors of amazing intricacy. How could anything make one or the other of them from scratch? Life had to come from simpler building blocks.

The challenge was that no one knew how RNA, or any other basic cell entity, might have arisen. A possible answer came from the deep.

Lurches and false starts

In 1943, when Günter Wächtershäuser was thirteen years old, his parents escaped from the Allied bombing raids on Munich to the small family-farm town of Giessen, north of Frankfurt. It was rolling agricultural country, with folklore and a chemistry tradition: nitrogen fertilizer had been invented there. The young Wächtershäuser studied and thought about the chemical world on a level beyond the bombs and the grim war news. He loved to think about big questions. Like Sidney Altman in Montreal, he cherished the day his chemistry teacher showed their class the periodic table. "There were only eighty elements in the whole world." He headed home to tell his father. "If that's true, I thought, then you could understand everything!" he said.

For years the challenge in finding the origin of life was that no one knew how RNA—or any other basic cell product or process, for that matter—arose from the amino acids Miller had created. Wächtershäuser earned a doctorate in chemistry at the University of Marburg but felt stymied by the conservativism of German science. Argumentative and driven, he left science for law school. In Munich he built a patent-law practice, specializing in science and showing why technical processes or inventions were truly novel. Bored as an attorney, however, Wächtershäuser kept his eye on origin-of-life studies, learning of the problems with Stanley Miller's theory when visiting with the Midwestern researchers Carl Woese and George Fox, to whom he was related by marriage. He became close to the British-Austrian professor Karl Popper, one of the main philosophers of the twentieth century, who mused on the science laziness that had frustrated Wächtershäuser as a doctoral student.

Popper made a huge impact on the young attorney, offering a critical insight that "destroyed the philosophy . . . which held sway over science for hundreds of years," Wächtershäuser later wrote. Supposedly science always moves from the particular, or experiments, to the general, or theories. But experimenters only see what they theorize, Popper said. Science thus resembles a carriage pulled by one blind, strong horse—the experimenter—and one weak, smart horse—the theoretician. The carriage lurches and makes wrong turns because of a constant battle between the two, Popper suggested.

With that new understanding, that science is not a perfect gradual progression toward truth, and with Popper's personal encouragement, the young attorney took up the study of the origin of life in earnest. He started with Darwin, who wrote in an 1875 letter that if life should be originating today in some remote part of the world, it would be consumed. "But if (and oh! What a big if!)" Darwin mused, "we could conceive in some warm little pond, with all sorts of ammonia and phosphoric salts, light, heat, electricity, etc. at the present day such matter would be instantly devoured." His notion sounded exactly like the Miller-Urey experiment.

The problem was, Earth's early atmosphere was probably nothing like that. It was mostly nitrogen and carbon dioxide. In Munich Wächtershäuser heard there were other problems with the Miller-Urey theory. The amino acids it produced would have been destroyed by the cyanide it produced. Furthermore, Miller-Urey could get nowhere near sugars, the backbone of RNA and DNA. Without genetics, there was no life.

Thinking about life's origin, Wächtershäuser started with the idea that life's chemistry could not change. The chemical laws that govern a cell's metabolism and respiration—the inventions of microbes—were laws. Life could not happen by some miraculous accident. It had to arise by law when certain geochemical conditions were met.

Wächtershäuser eagerly set out to understand what those conditions were. He set his mind to work during free moments in the courtroom or the office, scribbling notes on the train and at home in bed. By 1987 he realized that in the chemical pathways, where one chemical turns into another compound, "variations are not possible!" The basic metabolic pathways in life's central metabolism must be predetermined. "If that is the case," he realized, "then the life problem could be solved."

His heart raced, and his wife, Dorothy Gray, went out and got him a pile of notebooks. A historian, she understood his intellectual mania.

Fool's gold

By the late 1990s the origin-of-life field, if it could be called a field, was divided into three camps: those who believed that RNA or genetics came first, those who believed that metabolism came first, and those who felt the cell walls came first. Italy's Pier Luigi Luisi believed in cell walls first. Cech believed in genetics first. Those who were still struggling with the metabolism-first idea essentially followed the Miller-Urey soup theory. But not for long.

In Munich Wächtershäuser wrote page after page, snatching time on buses and trains and on weekends, all while running his law office. "It was imperfect," he recalled of his return to science. But "it was working!" He knew the idea of life's having originated in a shallow pond was wrong. The problem for the soup theory was that the movement of molecules in a pond was too random. How could trace chemicals floating in a vast primordial ocean find each other, link up, await a bolt of lightning, and then start metabolizing? It would take a miracle, and miracles "are not science."

Wächtershäuser became intrigued by two-dimensional surfaces under high pressures on minerals where chemicals might be coaxed to sit, contact each other, and congeal. Most organic compounds have a negative charge. If a mineral surface was positive, it would draw organics together, say under high pressure at the bottom of the ocean beside a volcanic vent, with its heat and constant eruption of gases. He focused on pyrite—commonly known as fool's gold—a mineral common at newly discovered deep-ocean hydrothermal vents.

In 1988 Wächtershäuser published, in something of a rush at the behest of Woese and Popper, his first draft of the idea. "It was not optimal," he recalled. "I made mistakes." He argued that no one had ever seen a flash of lighting ignite a chemical reaction. Rather, the origin of life had to proceed by formula from the Earth's geochemistry. The only place he could see that happening was at volcanic vents in the ocean. Wächtershäuser's first science publication in twenty years exploded into a field, causing tremendous anger and interest and even landing him a profile in *Scientific American*. The article, wrote the biologist David Deamer, "turned the entire concept of life's beginning upside down."

Wächtershäuser proposed a four-step process. First came carbon, in the form of carbon monoxide, which could come from the burning of any substance containing carbon. Then came volcanic gases, "undoubtedly available, by the laws of geochemistry." These offered hydrogen sulfide or cyanide. Then take transition-metal catalysts—iron, cobalt, and nickel—contained in the Earth's crust. Electrons from the volcanic gases react with the metal catalysts using cyanide, a process common in metallurgy and mining. "Volcanic gases," he suggested, "come in contact with metal catalysts like iron, cobalt, and nickel, in the crust of the Earth. These catalysts, with nutrients in the volcanic gases, form organic products which in turn, interact."

It was an impressive argument, but it needed Popper's formal tests. "It was not easy to figure out a way to prove my idea!" Wächters-häuser said. He took to experimenting with a collaborator, Claudia Huber, at the Technical University of Munich. In a series of vividly argued *Science* papers after 1997 that landed him in the *New York Times*, he showed how you could produce such microbial building blocks as the carbon-carbon bond to make the fundamental cell compound acetic acid. Two follow-up experiments published in *Science* and featured again in the *New York Times* showed that acetic acid could form amino acids, and that they in turn formed peptide bonds to make bigger organic molecules.

From the courtroom he had developed a polemical writing style, and he delivered his ideas in biting aphorisms. He called his idea the "iron sulfur world." It surpassed the Stanley Miller "soup theory," as he dubbed it, which, "with every modification," Wächtershäuser wrote, "increased in vagueness and ambiguity." In Miller's primordial pond or ocean, for instance, the molecules of life floated around randomly, with no impetus to connect. Instead, early life had to evolve on a two-dimensional, porous rock surface that provided protective crannies in which to concentrate and connect molecules close to where volcanic gases dissolved in the water and contacted the crust at the hydrothermal vents. "Catalysts come in contact with nutrients, forming organic products," he said. "These reflect back on or become attached to the catalysts, and modify the catalysts. Then you have better catalysts for eliciting new reactions."

The process actualized the potential present in an environment. He called it "a self-liberation process, which creates its own . . . unfolding of possibilities."

As Wächtershäuser was thinking and publishing papers, there came the earthshaking discoveries of teeming communities of life uncovered in the hot pools of Yellowstone and at volcanic vents at the bottom of the ocean. Because those strange microbes thrived in near-boiling water, eating rocks or acids, they were called extremophiles. Some fed on carbon dioxide, others on hydrogen sulfide, much as Wächtershäuser had predicted. But they had highly stable RNA in their ribosomes, or protein factories, and a number of researchers became interested in them for their extreme temperature tolerance to use as DNA replicators at crime scenes. Then came the discovery of springs in Kamchatka, Russia, where scientists sought to test Wächtershäuser's ideas in a natural setting.

If the modified catalyst makes more of the same, that was reproduction, Wächtershäuser thought. "If it makes something else, that is evolution. Life's origin was a process of self-liberation," he said to me. "When you have a soup, it limits the number of possibilities, of solidification from chaos." Life began in two dimensions, with restricted possibilities. Then life frees itself by conquering more space.

It ended with a "peculiar cosmological outlook," Wächtershäuser admitted. In the entire universe, life may have only one way of getting started, and anywhere those conditions were met, it would start "anywhere, anytime, in the same unique way." Origin was not in a time; it was in a place. Evolution was a process of conquering new places.

Some scientists latched onto it, but with the theorists it was not a popular idea. "Anyone who thinks life started at the bottom of the ocean is a fool," said Stanley Miller. But the influential pyrite theory and the RNA world hypothesis both showed that life must be predetermined by the laws of chemistry, argued Wächtershäuser, who never forgot about his high school teacher telling him the whole world consisted of some eighty elements. It seemed as though the next step had been taken.

The extreme microbes excited great interest for their tolerance of very high temperatures and for their potential to bioremediate toxic sites, and even as new energy sources. The field was reignited. After the failure of attempts to capitalize on Miller's 1952 experiment, however, Miller and his colleagues insisted life could not have started at the ocean bottom, arguing that organic material could not withstand such pressure. Miller's colleagues Jeffrey Bada and Antonio Lazcano went further. They argued that the organic material in seawater would fall apart. At Florida, the young astrophysicist Steven Benner claimed that water would destroy the pre-RNA nucleotides.

These two discoveries—that RNA, thought to be only a messenger, could build proteins and catalyze reactions as well as store the information that allows life to copy itself, and that teeming communities of life at hot vents on the ocean bottom might rival in size and diversity the visible organisms on the surface—offered further clues to the perplexing mysteries of where life came from. One insight was coming into focus, though: extreme microbes were found to live everywhere. Like creatures out of science fiction, they could ingest volcanic gases, cyanide, radioactive uranium, and various minerals. They could live in the Earth's most extreme and toxic environments, and some had been thriving since the Earth's beginning.

Years later those extreme microbes became hotbeds of interest for their industrial potential in high-temperature and high-pressure machinery. An enzyme in one of them, from the Earth's surface, gave police the key to DNA forensic analysis. In retirement in his home in North Carolina with his wife, Wächtershäuser was not thinking of those applications. Rather, he rather held on to the revelation he had experienced as a teenager, rushing home to his father from his first day in high school chemistry. "The world is orderly!" he announced to his father. "You can understand everything!"

Return of the Ancient Ones

In 1996 the twenty-five-year-old Susan Barns was walking along the slimy black sand pool called Obsidian Pool in the hills of Yellowstone National Park, trying not to touch the soft, muddy sulfurous mats around its rim. Blue, brown, turquoise, or deep aquamarine, the microbial films clung to the black rocks in the mountain valley far from the tourist-packed paths. It was a dangerous place to walk. You needed bear spray or bear bells, and coyotes often howled in the distance. The rocks were slippery; with one misstep she might plunge into a simmering cauldron tens of thousands of years old. She kept her eyes on the edge, scanning for the bison tracks that would tell her it was safe to step.

Barns was a former New York University art student who had found the art world too glitzy and volatile for her taste. Her father was a scientist, and she loved nature, especially the molecular biology of the extreme microorganisms that were being discovered in volcanic vents, at the ocean bottom, within mines deep inside the Earth, and on frozen mountaintops. She had wanted to get involved, and she had a skill that researchers of these life forms needed: she knew how to photograph cyanobacteria, the ancient, ubiquitous oxygen-producing microbes that produced Yellowstone's beautiful Grand Prismatic Spring. Barns contacted the discoverers, who needed help to investigate clues about these microbes' medical, environmental, and industrial potential. She was thrilled when they asked her to join.

The University of Illinois's pioneer theoretician, the evolutionist Carl Woese, pieced together the categories of primitive cell structures in his quest to understand the history of life. A Miles Davis poster and an American flag hung in his darkened office in Champaign, where he studied giant transparencies of ribosomal RNA. This RNA molecule of the cell protein factories "was the bar code of an organism." Because the methane-producing microbes seemed so unusual, he obtained microbes from such odd sources as a cow's gut and sewage sludge. He spent long hours staring at magnified transparencies in front of a light box in his cubicle. Others were pioneering new DNA technologies to unlock the human genome, but Woese was using a cutting-edge, painstakingly difficult RNA technology to decipher the origin of obscure microbes in a quest for the history of the cell. "It was to him the biggest quest, almost religious," observed the Woods Hole Laboratory Director of Microbiology Mitchell Sogin. "He treated science as some sort of holy temple. He worshipped truth."

Some of those microbes were closely related to those found in the Yellowstone pools, at hot vents beneath the ocean, and inside the human body. Woese saw something odd: their structures shared critical characteristics with plant and animal cells, but not with the bacteria they were supposed to be. Funded by a small NASA grant for seeking extraterrestrial life, Woese had discovered a bridge between one-celled and multicellular life. In those pools and in the guts of many animals, including people. lived a newly discovered third form of life, which he called Archaea, meaning "the ancient ones." Instead of the standard two kingdoms of life, bacteria and everything else, there were three — bacteria, Archaea, and everything else. Woese proposed a whole new vision of life's diversity and scope, as well as clues to its origin.

That controversial discovery ignited science. Because Woese avoided conferences and few understood how to read his transparencies, he was viewed as a crank whose method was suspect. He had based his whole claim on one ribosomal RNA gene. Stalwart microbiologists objected. Of course the microbes were just unusual bacteria, they said. The more NASA trumpeted the claim, the more it provoked a backlash.

Eventually, however, Woese and his postdoctoral researcher George Fox ushered in what *Science* called a "new era in one of biology's grandest, if most problematic, pursuits: understanding the origins of life." They had found a whole "new kind of cell in a teeming universe of microorganisms from primordial Earth that lived in oddball and overlooked environments, hot vents and swamps," offering "a new paradigm for understanding life on Earth." From there the race was on to identify more of these extreme microbes. What followed transformed our understanding of fundamental and applied biology.

Thinking hard, delicately balanced beside the Obsidian Pool, Susan Barns bent to retrieve a pan of mud from the simmering water.

"Those aren't bacteria!"

In the early 1980s Woese was among the first to see that visible plants and animals make up only a small part of the living world. Today we would go even further: bacteria and Archaea dominate the Earth's biosphere, and every other visible organism is an amalgam or offspring of those two.

George Fox came to Champaign, Illinois, after attending Syracuse University as a chemical engineer. He read Woese's article on protein evolution, a topic that was exactly what he wanted to study. He thought that clues to the origin of life lay in the cell protein factory called the ribosome. With Woese, he used primitive computer cards to analyze the painstaking crude RNA slides of every single known organism, setting the new standard for species identification with their 16S rRNA method. But Woese was a loner with mild Asperger's syndrome who went home every night saying to himself, "Woese, you have destroyed your mind again today." It was a heroic effort to catalog tiny spots, year after year, assembling the result into a more accurate tree of life.

The new Woese tree recognized the dominance of diverse, virtually ubiquitous microbes over all other life forms. Microorganisms defined the world: without cyanobacteria there would be no plants and without mitochondria there would be no animals. Much of the world's total biomass was microbial. If you wanted to understand climate change, human health, or the planet's future, you had to understand the microbial underpinning that "keeps the biosphere stable." As new methane-producing Archaea were found at the sulfuric hot vents at the ocean bottom, then floating in the oceans at large, then in us, Woese and Fox proposed that the new tree of life should not be considered a tree at all, but rather a series of radiating spokes of misshapen wheels.

All the microbes needed to survive, it seemed, were volcanoes, metals, water, and hydrogen. Then it was seen that Archaea could live almost anywhere — in the ocean, in soil, even inside of us. Certain strains could help clean up toxic waste and generate new genetic tools. They engaged in gene swapping and exponentially increased the diversity of the microbial world. They had industrial uses, energy-generating possibilities, and remediation capabilities. They and their bacterial cousins recycled everything that died into soil or minerals or gas, regulated key climate-warming gases such as carbon dioxide, and tapped the Earth's energy. Their chemistry offered a blueprint for confronting the energy crisis, perhaps generating hydrogen from sunlight, biofuel from plants, recycling waste into food, and switching from methane to hydrogen fuel cells. They affected global warming in ways we did not yet understand.

Still, we knew little of their true diversity or makeup. "It was like saying you want to study animals and not being able to distinguish a rhinoceros from a plant," Woese told me. Who were these strange creatures, and what could they do?

Most of what we knew stemmed from microbes that caused disease or spoiled food or milk. Few people were interested in strange microbes from extreme environments, even if they had industrial uses, until someone actually showed how it could be done. The big money potential of the newly discovered species became clear when two researchers from Canada and California got involved.

Crime scene investigation

Thomas Brock was an unlikely biocapitalist. The mild Canadian-born researcher studied the cyanobacteria in Yellowstone's hot pools as a way of understanding the evolution of photosynthesis. The origin of this most amazing life ability—to take the energy from the sun, using quantum mechanics, and separate the hydrogen and oxygen from dirty water, burn the hydrogen, and exhale the oxygen—fascinated him. To understand how very early, primitive, extreme microbes ever developed this skill in primordial Earth, Brock took a keen interest in Octopus Spring, where pink filamentous bacteria, which lacked the ability to photosynthesize, seemed to live at the unheard-of temperature of 170 degrees Fahrenheit (almost 80 degrees Celsius). He could not cultivate these microbes, so he switched to the somewhat cooler organisms from Mushroom Spring, which lived at about 140 degrees Fahrenheit (60 degrees Celsius). After years of work in the 1980s, he managed to grow them in his lab. Studying their ribosomes, or protein factories, he realized it was a brand-new, heat-tolerant microbe and named it *Thermus aquaticus*, source of the heat-resistant enzyme *Taq*). He later found that *T. aquaticus* lived in hot-water taps around the world, even at the boiling point.

The fact that *Taq* had the ability to replicate DNA quickly under extremely hot conditions interested the Berkeley–based Cetus

Corporation became interested. It caught the attention of the Department of Energy and DARPA, the Department of Defense's intelligence program. But no one knew how to harness the enzyme's unique ability.

The California surfer and chemist Kary Mullis liked to synthesize LSD. He was driving along a lonely coast highway at night, thinking about the *Taq* enzyme, with his girlfriend sleeping beside him. He was considering the repetitive nature of the polymerase chain reaction of *Taq* when something occurred to him: if you let the polymerase go, it would set off a chain reaction, replicating DNA indefinitely. He pulled over and woke up his girlfriend, excitedly telling her his idea because he feared that if he did not, he would forget it.

Cetus Corporation soon developed the *Taq* polymerase for DNA kits used in crime-scene investigations. In 1991 the pharmaceutical giant Hoffman-La Roche bought the technology for $300 million. In the 1990s it was making $100 million a year from its kits. The age of microbial prospecting, sometimes illicit, had begun.

The extreme microorganisms had other industrial uses. Their oils could be used in high-temperature machinery. Their *Taq* enzyme became a key ingredient of lab DNA kits. Archaeal enzymes became a part of detergents, starches, and solvents, beer and wine fermentation processes, and the production of baked goods, paper, syrup, and juice. Their potential got the government's Energy and Defense departments even more interested.

Then researchers in the deep-sea submarine *Alvin* discovered giant communities of life at boiling springs at the ocean bottom below the Galapagos Islands, some at 300 degrees Fahrenheit (149 degrees Celsius) and some hotter still. The industrial and financial potential of their strange abilities was significant. The early microbes lived off of iron, hydrogen sulfide, and cyanide. The federal government was so interested that in the 1990s the Department of Energy established the Microbial Genome Initiative to fund the sequences of these and other extreme microbial genomes.

Yellowstone granted researchers dozens of permits to collect microbial samples in the hot springs, raising a huge controversy. "Never before has industry profited directly from living creatures taken from a National Park," wrote Brock, and critics accused Yellowstone Park officials of missing the boat on a windfall. The ethics would be discussed at length thereafter, as new companies negotiated more equitable national park deals for extremophile bioprospecting.

The Department of Energy was particularly interested in the fact that some of the cold and heat-tolerant archaeal microbes produce methane by the ton. Their oils could be used to protect machinery operating at extreme temperatures. Companies like New England Biolabs began selling archaeal vent and deep-vent polymerases as genomic research tools. Novozymes, DuPont, ArcticZymes, and Genencor were among the companies marketing the extremophile enzymes for various industrial uses.The Argonne National Laboratory went all-out to understand their uses in bioremediation, clean energy, and as pharmaceutical ingredients, as their lipids could be resistant to many biological disrupters. The next stage was to explore these industrial extremophile enzymes for bioremediation.

But to explore those, first one has to know what was out there. A Colorado researcher discovered how to do just that.

"Do it in the dirt"

Sitting in his lab on leave in Illinois, the fiery, diminutive Norman Pace was impatient. He was reading Brock's book on Yellowstone's Octopus Spring and learned that it held "kilogram quantities of pink filaments." Pace was an avid cave explorer who had nearly died in Chiapas, Mexico, when his anchor popped loose and he hurtled over a subterranean waterfall, losing his pack and glasses when, halfway down, his foot rope yanked him upside down, gasping, into the torrent. He was not the type to wait around. After reading Brock's book, suddenly Pace realized researchers did not have to cultivate microbes in the lab, which was exceedingly difficult. Rather, the thick rugs of brilliant pink bacteria in a Yellowstone pool could be picked up themselves in the mud. All you needed was their RNA, which could be transported in an ice chest in a van. "Let's go to Yellowstone and get a bucketful of phenotypes," he told his lab group.

"But you won't know what's in it," objected one postdoc.

"The RNA will tell you what you've got."

They sat there, stunned. "Do you know what you just said?" the postdoc recalled saying. "You transformed microbial ecology."

When he was deep in that cave in the mountains of Chiapas, the thirty-five-year-old biologist realized that caving was critical to his science, teaching "that if you think it's all known, you fall into a paradigm of self-satisfaction," he observed. "Yellowstone was like another cave to Norm, a kind of a pilgrimage," recalled David Stahl, who joined him on the first trip west. They stored the mud samples on dry ice in a metal cooler and brought the van to Illinois to start sequencing. The long-haired, bearded Woese held up Brock's book for a snapshot.

Directly pulling RNA from the mud offered a new way of seeing, more akin to life's reality than the old methods of Pasteur and Fleming. Microbes did not exist in isolation in the world; they lived in teeming symbiotic communities in which they borrowed freely from each other's food, waste, even genes. The strains of non-oxygen-breathing Archaea for instance, often lived with bacteria that did breathe oxygen. By the early 2000s, as the dozens of vent organisms in Yellowstone were also being found at the hot vents at the ocean bottom, living at the pressure of 300 Earth atmospheres, a race was on, funded by the Department of Energy, to sequence the genes of these unknown organisms.

Suddenly genome institutes were interested. The Bethesda-based Institute for Genomic Research (TIGR) leapt on the idea of extreme-microbe gene prospecting. By 1993 the institute, run by the genome researcher Craig Venter at the University of Maryland, was fighting the government for money to study the microbes. The Champaign and Bethesda groups scheduled a tense meeting to see who would control the Department of Energy funding. Woese suggested a variety of organisms to sequence, but Venter held the purse strings and did not want the data to leave the Maryland lab. Once the presentations were over, the groups went out for Chinese food. After a beer they loosened up and came to an agreement. "The motley crew did it," Woese said as they raced to make their plane home.

They began with *Methanococcus jannaschii*, recovered from three kilometers below the Pacific surface, and expanded from there. Some 56 percent of its genes were previously unknown, reported Carol Bult of TIGR. They kept going. As each microbe proved stranger than the last, from Iceland to Italy, it became clear that oxygen- or carbon dioxide–based organisms—animals and plants—made for only a small portion of life's full reach. That was where Susan Barns came in.

She took a special interest in Yellowstone's obscure twenty-seven-foot Obsidian Pool, its water boiling and spilling into lower pools. Barns sampled the DNA in the mud and found a new pair of Archaea that looked like the most primitive ones ever seen. By the time she finished, she had found a total of thirty-eight new species of Archaea, as different as cyanobacteria are from spirochetes, in one pool. Few researchers ever found even one new species. Most researchers would make a career out of discovering one new organism. Barns found thirty-one.

Then another postdoc, the Australian Philip Hugenholtz, took the same mud and found an even greater number of bacterial species—about fifty—than there were species of Archaea. The diversity of life was orders of magnitude greater than imagined. Eighty- eight new species had been found in one pool.

But the discoveries moved research onto something bigger: the origin of all life.

Look small to get the big picture

The impetus for the discovery of the new archaeal kingdom had been ribosomes, those cell structures that make proteins, and probably the transition structure from the early RNA world to the modern cellular world. Being such an old and vital structure, dating back to the very first life form, ribosomes had changed little over billions of years. George Fox went on to study the ribosome at University of Houston, proposing a timeline for the development of the protein factory fundamental to life. "We can look at the ribosome and see the parts that are more ancient," he said to me at a NASA conference in Chicago. "That last common ancestor had ribosomes! So by looking at the history of the ribosome, we are actually walking into the past before that last common ancestor."

In seeking to understand the ways in which that ribosome evolved and its mechanism of change, Fox learned that RNA synthesized the protein pathway with no help from proteins. It worked almost like a loom. At the ribosome center, where the genetic information transfer occurred, "there's absolutely no protein anywhere!" he told me. Together the Archaea and ribosome studies offered deeper glimpses into the symbiotic relationships underlying early life.

The Israeli biologist and Nobel Prize winner Ada Yonath explained why that is. Raised in a poor family in Tel Aviv, inspired by the career of Marie Curie, Yonath made the first beautiful, intricate photographs of the structure of the ribosome. Her sickly father died when she was a child, and Yonath worked outside the home to help her mother and family. At eighteen she completed her compulsory army service in the top-secret office of the Medical Forces. Yonath completed her undergraduate and master's degrees in chemistry, biochemistry, and biophysics at the Hebrew University of Jerusalem. At the Weizmann Institute she attempted to reveal the structure of collagen, continuing her work in her postdoc at the Mellon Institute in Pittsburgh and at MIT. At the end of the 1970s Yonath returned to the Weizmann Institute, where she started Israel's first biological crystallography laboratory in order to study ribosomes.

Yonath's major research involved protein synthesis, which she hoped to determine through a three-dimensional structure of the ribosome. To reveal the three-dimensional structure, crystals were required, but dealing with ribosomes presented additional challenges. "The ribosome is a complex of proteins and RNA chains; its structure is extraordinarily intricate; it is unusually flexible, unstable and lacks internal symmetry, all making crystallization an extremely formidable task," Yonath wrote later.

In the 1980s Yonath's inspiration came from reading an article about hibernating bears, which "pack their ribosomes in an orderly way in their cells just before hibernation, and these stay intact and potentially functional for months," she recalled. She assumed this was a natural strategy for organisms in harsh conditions. By taking ribosomes from the bacteria living in environments such as the Dead Sea, thermal springs like those in Yellowstone Park, and even atomic-waste piles, Yonath and her team produced the world's first ribosome microcrystals in work often mocked by other scientists for its arcane purpose and deep difficulty.

Her lab's map of the ribosome's small subunit made a breakthrough. In 2000 and 2009 her team published the first "complete three-dimensional structures of both subunits of the bacterial ribosome." This structure made a long, gated tunnel through which a protein "progresses as it is being formed," she said in her 2009 Nobel acceptance speech.

The significance is that this complicated structure, comprising RNA, enzymes, and metabolic machinery, was also present in the very first living cell, or LUCA (last universal common ancestor), shortly after the Earth's formation. The ribosome is the same in virtually all life, from primitive *E. coli* to the most highly evolved animal. By understanding the ribosome, Yonath paved the way for the structural design of better drugs to confront the resistant pathogens of our time.

The Weizmann Institute itself was a product of one of the first modern industrial applications of the power of microbes. Early in World War I Britain's young undersecretary of the Navy, Winston Churchill, was desperately seeking a way to make cordite, or smokeless gunpowder, at a time when Britain's cannons were reduced to shooting only four shells a day owing to its scarcity. Churchill put out a circular to scientists soliciting ideas. The University of Manchester biochemist Chaim Weizmann, then known as Charles, said he could make acetone by fermenting a strain of *Clostridium*. Churchill latched onto the idea. How much did he have? the undersecretary asked. He needed thirty thousand tons, to which Weizmann famously replied, "I have a few cc's in my lab."

The government commissioned six whiskey distilleries for the wartime microbe effort, including one in Toronto, and the U.S. government built two in Indiana. As a reward Weizmann, a committed Zionist, won support for the British Balfour declaration, as well as funding for the highly regarded institute in Israel. The *Clostridium* technology was commercialized in Peoria, Illinois, by a company called Commercial Solvents, which in turn took the waste butanol the process produced a version of the prized paint coating we call lacquer. The modern industrial version of an age-old microbe economy—that of wine, beer, cheese, and yogurt—was born.

By the early 2000s several labs had capitalized on Yonath's and Fox's work with better and better digital tools tracing the origin of the ribosome and of microbes in general. In so doing they raised a thorny issue, almost a deal breaker, for the tree of life that Woese had rewritten.

"Early life was not chaste"

In Earth's earliest years as a molten fireball battered beneath a distant sun, with freezing temperatures in some elevations, no ozone layer to protect its budding life and very little land, rent by volcanoes and meteors and greenhouse gases, the earliest premicrobial life could not have been considered as separate, distinct species. They shared genes freely in a kind of floating giant mass, clinging to their metabolic life. "Early life," quipped the Mexican researcher Antonio Lazcano, "was not chaste."

If they shared genes freely, even to the present day, then any attempt to trace their history and lineage was vastly more complicated than supposed. Symbiosis researchers such as Lynn Margulis, who noted how easy it is to move a gene from one animal to another, had foreseen the problem. Gene swapping challenged the whole notion of categorizing the most ancient microbes. The tree of life was withering.

A series of computer simulations by the National Institutes of Health's Eugene Koonin and the University of Illinois's Nigel Goldenfeld provided some direction to the final quest of Carl Woese's life: to understand the logical processes behind the evolution of life's information program. Before genes there were progenotes, early recorders of genetic information. But why were they in the form of nucleic acid? Why are there three forms of life and not more? Why were there a durable library (DNA), a fragile messenger (RNA), and intricately designed workers (proteins), that must fold properly to succeed? Would the problem of swapping mean there was no way to reach back to life's origin?

It turned out that if you used the latest in computer information theory, the most optimal way to organize a living system was just the way it had been organized. Life's genetic code, said Goldenfeld, a short-statured South African mathematician, was a one-in-a million information system to assure accuracy. It resembled the algorithm on a computer. "Roughly three billion years ago, microbial life invented the Internet," Goldenfeld explained. "The information is stored in molecules. . . . Want to know how to become a more virulent pathogen? Download the gene!" The significance of the metaphor was that the tree of life had held, at least in the form of a meme or model.

The history of life was a history of processes rather than of finished products. Evolution was not a "progression of forms," concluded Carl Woese, but a "progression of processes." Early life was collective, and the core cellular machinery, such as protein translation, was horizontally transferred, perhaps much as Margulis proposed. With Goldenfeld Woese wrote one final paper trying to understand the computational rules by which those processes developed.

Open your refrigerator or pick up a handful of soil. You are looking at billions of previously unknown organisms, most of which we still have not identified. Some could live miles below the Earth's surface, some in ice, some in the depths of space, some in boiling water at the ocean's bottom. Yet they contain vital clues to our health and our planet's future. A new finding at the bottom of the ocean suggested how.

Go storm a castle

While some cell organelles, like the mitochondria, can be traced back to bacteria, the machinery for processing genetic information of plants and animals is more akin to that of Archaea. We may have descended from a common ancestor, or from a union of Archaea and bacteria. But there is now evidence that plants and animals emerged directly from Archaea.

In 2015 the Norwegian scientist Thijs Etemma and others made a startling discovery at a gigantic field of stone castles lying at the bottom of the North Sea. They had been working at the remote site for years. Thinking there could be evidence of life's lineage in the genomes of microbes at its base, the authors worked out the base genes from marine sediments, using sophisticated tools of digital investigation. After years of struggle they put together one complete genome and two partial genomes of closely related members of the Deep Sea Archaeal Group. With the help of marker genes that remain over long periods of evolutionary change, the researchers connected this new archaeal group with eukaryotes, or complex life, thus establishing the new group as the nearest relatives of plants and animals. The group of Archaea was called Loki, or Lokiarchaeota, because they live by Loki's Castle, a field of towering, gothic hydrothermal vents in the mid-Atlantic named for the Norse god.

The Loki genomes were a group of similar extremophiles. Some 32 percent of the proteins coded for by Loki genes are unique, with no known precedents. About 26–29 percent show affinity to archaeal and bacterial proteins, suggested there has been gene exchange between Archaea and bacteria. But a large number, 175, or some 3 percent, of the proteins were similar to those of eukaryote proteins, mainly those involved in membranes. The Lokiarchaeota thus implied the emergence of key plant and animal proteins even before the bacterial features of mitochondria appeared. They seemed to represent a "starter kit" for "support[ing] the increase in the cellular and genomic complexity that is characteristic of eukaryotes," Etemma told *Nature*.

In 2016 a new deep-sea and Yellowstone discovery offered further support for the idea that the origin of complex life happened when an archaeon swallowed a bacterium. The evidence came from newly discovered organisms much like Loki, named for the Norse gods Thor and Odin. "Like Loki, these Archaea have a lot of the proteins that make a eukaryote," said the University of Texas's Brett Baker, who helped make the discovery. They offered further evidence that the microbial dark matter—all the unknown genetics of microbes—supported the controversial claim that all plants and animals, which is to say almost all visible life, were merely an outgrowth of Archaea. By 2017 a combined team had found a whole new supergroup of similar primitive microorganisms that they named for another Norse god, Asgard, and that contained precursors to many proteins found in complex cells. Asgard lived in the Lost City hydrothermal fields, but also in mining-waste runoff near Rifle, Colorado, and in a placid North Carolina river, in New Zealand, in Japan, in the hot springs of Yellowstone, and quite probably all over the world.

Discoveries like these deepened the mystery of where we came from, while the strange abilities of the new microbial families offered clues to understanding "how life functions on the planet," Woese once said to me. From there, it became possible to glimpse the potential of synthetic biology, or designing microbes to try to address climate change or provide new energy sources or, even, new approaches to medicine.

Researchers such as Fox, Barns, Woese, and Pace revolutionized the understanding of life's breadth and thus the search for its origin, providing definitive evidence of symbiosis in the evolution of plants and animals and bringing microbes to the central dominating role in the tree of life they deserve. The Archaea took hydrogen and carbon dioxide and made methane. By finding and studying a whole new kingdom of previously unknown microbes, they paved the way for a new understanding that transformed our vision of ecosystems, climate change, and life's origin. "What's really interesting is not what's going on now," Woese said to me. "It is what's going to happen."

The vast majority of life on earth was microbial, and the vast majority of microbial identities were not known. The accepted estimate was that 99 percent of microbial species in a given environment were still unknown. Over the years a global effort was made to decipher more of the dark matter by means of learning their genome DNA sequence from environmental samples. The idea occurred to a number of researchers that some of these newly discovered creatures might lead to new antibiotics. They certainly gave a clue to the origin of complex life. The extremophiles thrived in boiling or freezing water. Up to thirty companies, including startups like a San Diego-based company named Diversa as well as giants like ExxonMobil, obtained permits to mine Yellowstone hot pools for the extremophile enzymes with industrial potential. Some may offer clues to new approaches to medical treatments.

The challenge was huge. In using the newer field DNA sequencing technologies, microbial diversity proved to be staggering—greater in sheer number than the stars in the universe. Pick up a handful of soil or a glassful of seawater. You are looking at billions of organisms more diverse and hardy than ever imagined, perhaps 35,000 species in soil, fewer in seawater. They produced the oxygen we breathe. Imagine if we could harness their powers.

At the same time, the origin-of-life search once again hit an impasse. By 1995 many in the search were depressed. "Miller writes a paper saying there's no way of getting ribose, sugar. This whole thing is going to collapse," commented Florida's Steven Benner. "I last saw Miller in his wheelchair in 2003. 'The field has not made any advances,' he complained to me."

Extreme microbes, a third form of life as ancient as the first, offered a way out. Amino acids that formed protocells and a new understanding that our body units were comprised of multiple free-standing bacteria together set the stage for a revolution in understanding. "The Woese paper is one of the most influential in all microbiology," the Stanford University researcher Justin Sonnenburg told a reporter. "It ranks with the works of Watson and Crick and Darwin." Its ideas caught fire with many of the younger new generation of microbe hunters. "It is the single most important paper in the history of microbiology," said the University of Chicago's Jack Gilbert, editor of a new journal on microbial science called *mSystems*. "He laid the foundation for the visionary statements about microbial ecology that basically formulated the next thirty years of the field."

But what I recall most are comments Woese made to me about personality and science. "Science does not belong to the scientists," he said. "You have to have your own particular sensitivity to the world. And there are parts of it that are beautiful to you no matter what anybody else thinks."

Ribosomes looked like the critical transition structures from the early RNA world to the modern cellular world. It became possible, using DNA sequencing of the extreme microbes in the most inhospitable regions of the Earth, to glimpse which of the parts in the tunnel-like cell structure were the oldest. Together the Archaea and ribosome studies offered tantalizing glimpses into the symbiotic relationships underlying early life. Two grand problems of microbial ecology became the origin of the cell and the future of the global environment. What only a few realized was how closely the two were linked.

Then, about the time Woese began looking at strange microbes from a cow's gut, two meteors hit the Earth.

5

Shooting Stars

The September mid-morning sky over Murchison, Australia, was clear and blue, spattered with the shards of clouds over the farming city grid and echoing with the rattle of delivery trucks. Suddenly a light flashed in the sky over a strip mall, growing larger and larger. It appeared to be headed right past the Murchison River into the surrounding desert. At the last second the bright light burst and separated into three brilliant yellow fireballs, each hurtling in a shower of terrifying sparks moving at twice the speed of a bullet. Thirty seconds later a tremendous boom sounded, and smoke rose from several points in a clearing in a field. Pieces of smoking boulders and rock scattered over some five miles of Australian farmland. One barn was on fire, and people hurried out of their houses to put it out.

It took several days for scientists to piece together what had happened. Not until October 9, 1969, did the Australian Astronomical Society issue a bulletin on the fall of a meteorite over Murchison. Researchers from around the world booked flights to get in on a hunt that yielded more than 220 pounds of a rare primordial rock.

The year 1969 was a good one for space exploration. It featured the Apollo 11 moon landing and two spectacular meteorite strikes in Mexico and Murchison. In February a meteorite the size of a Buick raced across the skies over the Mexican state of Chihuahua, breaking into pieces over the small town of Pueblito de Allende. Bigger than Murchison, it resembled its Australian cousin as a rare form of meteorite that contained within it the building blocks of life. Some suspected these pockmarked meteorites, called carbonaceous chondrites, brought the amino acids that formed life on Earth. Because they originated in the 4.57-billion-year-old dust cloud that formed the solar system, few had ever been analyzed.

Every time I visit Chicago's Field Museum I stop on the second floor to stare at the gnarled black shards of Murchison and Allende. Those two meteors shot across the sky with a sudden roar that shattered the desert stillness, splattering remote mountains with thousands of pounds of smoking, tarlike rock. Scientists raced to get pieces, as did locals and scavengers who nabbed them to sell on the black market.

In Tempe, Arizona State University researchers scrambled into action. John Cronin and George Yuen, co-directors of the university's Center for Meteorite Collection, had painstakingly built up the world's largest meteor collection in Tempe, not far from the Arizona Great Crater. They had to get hold of Murchison because the bits and shards of rock were so rare, formed when asteroids smashed into each other. The researchers hoped the debris would harbor organics, including the twenty amino acids that form the components of proteins. Once they got the rock they needed help to analyze it, preferably from someone reliable and detail oriented.

Both Murchison and Allende proved to be rare specimens, among the 4 percent of precious meteorites containing abundant carbons. The materials of life, such meteorites suggested, float in the vast, icy stardust of space, in giant clouds with tons of icy crystals and organic compounds. Those building blocks, it turned out, shared certain critical structural qualities with Earth life that seemed hardly to be coincidence. A revolution in technology beginning in the 1990s offered to reveal the dawn of life in the solar system. But it took an obscure Italian woman and mother of four to help get it started.

In Padua

Born in 1932 and raised in wartime Venice, Sandra Pizzarello studied ancient Greek for four years and Latin for eight in her city public high school. She went on to a doctorate in biochemistry from the school in Padua where Galileo taught mathematics. After immigrating with her husband to Tempe, Arizona, in 1970, she worked for an Italian pharmaceutical company and raised four children. Once they were all in school she decided to take a graduate course at Arizona State. At age forty-three she wandered into a chemistry class and thought her mind "would explode," she told me over lunch at the Chicago Hilton.

She caught the attention of two professors, John Cronin and George Yuen, who had a NASA grant to study the compounds found in meteorites. Meteors are fiery shards of the asteroids formed in the dawn of the solar system, and meteorites are those that fall to Earth. Cronin offered her a job analyzing the Murchison pieces the university had obtained. In the morning Pizzarello sent a daughter off to her elementary school and then went to work. She put the shards through an ionizer and found the amino acids they contained had never been seen on Earth before. More important, their molecules favored a left-handed structure. Most of the amino acids of Earth life shared a left-handed structure, an anomaly that had long puzzled researchers into the origin of life.

Like your left and right hand, organic molecules can have the same elements in one of two mirrored orientations. Just as a left-hand glove can never fit a right hand, the fact that organic material in space shared the same structural orientation as Earth life suggested that meteors might have planted life's seed here. But Cronin was skeptical. A previous meteorite detection of left-handed organic molecules had been exposed as contamination.

Comets have long fascinated people. The ancients thought they were evil omens, and the Druids and others feared or worshipped them. In 1869 in France, the very first carbonaceous chondrite meteorite landed in Paris's seventh arrondissement on a cold spring morning, providing a possibility of studying meteors. In 1986 NASA's Giotto mission offered the world's first proof that comets are small icy rocks with abundant water. The field advanced rapidly with the discovery of amino acids on meteorites in 1990. But it was in 1997, when Pizzarello and Cronin published in *Science* their paper on the left-handedness of the Murchison meteor amino acids, that the field took off.

As Cronin feared, the paper ignited controversy. Some astronomers claimed their sample was contaminated. But Pizzarello had an ironclad response: the left-handed amino acids on Murchison had never been seen on Earth before—so the rock could not have been contaminated. The fact that comets like Murchison, carbonaceous chondrites, are fragments of primordial asteroids formed in the beginning of the solar system was even more compelling.

Further intriguing, poetically inflected evidence came when a chemical analysis of the four major compounds in Halley's Comet— carbon, hydrogen, oxygen, and nitrogen—proved to be in exactly the same proportion as in the human body. The astronomer David Levy, co-discoverer of Shoemaker-Levy, called it "a compelling hint" that meteors seeded the Earth with the ingredients of life.

At the same time, a race to uncover Earthlike planets outside the solar system was generating world interest as competing Swiss and American teams, using a technique for analyzing shivers in star spectra caused by the gravitational pull of their planets, uncovered planets orbiting other suns. The number of exoplanets rose from a trickle to a torrent, from dozens to hundreds to nearly a thousand. That achievement over a few years, with inexpensive equipment, brought new money into the search for life in space. Coupled with the glimpses of microbes at ocean bottoms, in the stratosphere, in mines far below the Earth, in the bone-dry Chilean desert, in the Great Salt Lake and the Dead Sea, and in permanently iced lakes of Antarctica, it seemed the conditions for microbes exist in many places in space and virtually everywhere on Earth. What that meant became a topic for heated debate.

Hot air

In April 2002, at Oxford's Balliol College, the white-haired Martin Brasier stepped before a crowd in conference to skewer a claim by California's Bill "Bull" Schopf that he had found fossils of bacteria in 3.5 billion-year-old remains in northern Australia. Then, in 2011, Brasier himself claimed that he had found the world's oldest bacterial fossils in the 3.4-billion-year-old Strelley Pool formation, only a few miles away from Schopf's.

In San Diego Jeffrey Bada, of the University of California, uncovered in a campus storage closet the remains of the original Stanley Miller experiments, included the apparatus and notes on hundreds of repeated tries Miller had performed after moving to UCSD. When Bada's students retested the contents of the yellowed boxes, they found many more amino acids than Miller had reported. Amino acids could be made pretty readily from simple building blocks.

An ongoing battle was also raging between the cell-walls-first proponents, the genetic-material-first camps or RNA world group, and the metabolism-first proponents. The question they wrestled with was, if RNA came first, then what was there to separate the nucleic acid from its surroundings? If walls came first, as in bubbles around a volcanic vent, then what were they walling off?

Then, in 2005, Sandra Pizzarello published in *Geochimica et*

Cosmichimica Acta the dating of deuterium enrichment of amino acids in carbonaceous chondrite meteors, meaning that their origin, the comets and the organic material on them, pre-dated that of the sun. Pizzarello also subjected the rocks to hydrothermal pressure, which they would have undergone if they had landed in the sea, unleashing a wider variety of organic compounds from the stone. At the same time a school of thought was gaining ground that Mars was the likely starting point of life; water there had been abundant and conditions calmer than on the very early Earth, which had just undergone a collision with another planet, vaporizing everything and leaving the moon in a close orbit around a gigantic, heated wreck. Fragments of Mars hit our world regularly and are most readily found in the vast white ice fields of Antarctica. These would have been more numerous, the theory went, in the solar system's earliest days, when meteors and asteroids regularly bombarded the five close-in rocky planets.

This roller coaster of findings in the early 2000s formed the backdrop for NASA's decision to send off a series of three dramatic probes of passing asteroids and comets. On July 4, 2005, NASA's Deep Impact spacecraft arrived at Comet Tempe 1 to impact it with an 820-pound instrument, leaving a crater the size of a building some twelve stories deep. Ice and dust debris flew up from Tempe's surface in a brilliant plume visible from Earth. Sunlight reflecting off the ejected material provided a dramatic brightening that faded slowly as the debris dissipated into space or back onto the comet.

Leading that effort was the astronomer Karen Meech, who as a young girl had loved Gene Roddenberry's television series *Star Trek*. Meech graduated from Rice University in 1981 with a bachelor of science in space physics and earned her Ph.D. in physics and earth and planetary science at the Massachusetts Institute of Technology, moving to a professorship at the University of Hawaii and investigations on three NASA missions—Deep Impact, EPOXI (Extrasolar Planet Observation and Deep Impact Extended Investigation), and Stardust. The coolest thing about Deep Impact, she later said, "was that we were actually going to perform an experiment. . . . Unlike much astronomy where you don't get . . . on site." A worldwide science community watched Deep Impact hit a comet surface, hoping that it would give clues to the primitive material inside comets.

There followed a series of daring and somewhat unlikely probes, including EPOXI, which in 2009 sent back vivid images of comet Hartley 2 before firing itself off into oblivion. But the biggest success came with the Stardust mission.

What dreams are made of

In 2007, when the NASA Stardust team presented its discoveries at the annual Lunar and Planetary Science conference only three months after the probe landed and sampled the rock and ice of the comet Wild 2, "you could see jaws drop in the room," recalled the principal investigator, Don Brownlee. The ice of Wild 2 was, as expected, from the frozen expanses a few degrees above absolute zero beyond Neptune. But the comet's rock had formed under white-hot conditions a few miles from the sun. The comet studies provided a direct look at the nature and origin of the materials sprayed all over the young solar system and incorporated into all its planets and moons.

The discovery initiated an upheaval in understanding. The early solar system was a nonstop pinball machine in a rapidly whirling disk with huge temperature and energy gradients, all of which combined to create the ingredients and processes that gave rise to life. Comets were made of mineral-rich deposits from that time before planets, seeded by stars much as we were. "We are stardust," Joni Mitchell sang in her song "Woodstock," giving NASA its name for the 2003–9 mission. Among Wild 2's high-temperature materials were odd, rounded particles called chondrules and white irregular particles known as calcium aluminum inclusions (CAIs). Chondrules are droplets of rocks that melted and then quickly cooled as they orbited the sun. Much rarer than chondrules, CAIs are the oldest solar system materials, exotic minerals that form at high temperature. The rocks forming the bulk of the comet's mass came from a zone hot enough to melt bricks, right by the sun.

The seventy-two images of Wild 2 were dramatic, said Brownlee: "kilometer-sized deep holes bounded by vertical and even overhanging cliffs," he wrote, with "flat topped hills surrounded by cliffs; spiky pinnacles hundreds of meters tall." What they did not see were craters like those found on the moon. The lack of craters indicated that the surface was fresh and new, perhaps a few hundred million years old, contradicting the idea that meteors date back to the solar system's origin. It was a mystery.

Stardust also had one last surprise: the 2009 discovery of the amino acid glycine by scientists from the Goddard Space Flight Center. That this molecule could be detected in the tiny particles collected at six times the speed of a bullet fired from a rifle was a technical triumph that incorporated the use of isotopic composition to prove that the glycine was not a contaminant from our own planet. The significance was that comets must have delivered at least one amino acid to our planet before it had life. Because most stars have comets, it suggested that any Earthlike planets obtain important prebiotic molecules from space. Comets carried a treasure trove of amino acids, some with the same left-handed structure as those in all living things on Earth.

The search for microbes on meteorites seemed to be on the verge of bearing fruit. Then came a major test of microbes' potential in the real world.

Evening light

On the evening of April 20, 2010, all was still on the 137-person BP oil rig fifty miles southeast of Louisiana. The waves rustled, and a light wind stirred. Then warning alarms sounded as oil gushed back up the main pipe.. Suddenly an explosion shattered the stillness as workers stumbled from their beds. Eleven of them died. For eight months 4.6

million pounds of oil gushed into the Gulf of Mexico, ultimately contaminating close to five hundred miles of coastline.

The Deepwater Horizon oil spill was the largest petroleum accident in history. Tens of thousands of volunteers and professional emergency workers raced to clean up the hundreds of miles of contaminated beaches and swamps and save wildlife. They noticed that at sea the oil slicks could be degraded if one added nitrogen, in the form of, for example, fertilizer, to the contaminated water to produce natural oil-eating microbes such as *Thassalitus oleiverans* and *Alcanivorax borkumensis*, mostly found in the Gulf of Mexico. Their lead researcher was the University of Georgia professor and microbial ecologist Samantha Joye.

Anaerobes made all of the world's oil, and various bacteria loved to eat it. The field of microbial bioremediation, the use of natural microbes that thrive in the most forbidding, high-pressure environments, to clean up oil, toxic waste, and even radioactive contamination was born of necessity and ingenuity. The ancient Romans used microbes to clean their famous sewer systems. During World War II the U.S. Navy cleaned up wartime oil spills with natural petroleum-eating microbes. In the 1970s General Electric's Ananda Chakrabarty developed several oil-eating bacteria to clean up oil spills. Auto repair shops pour commercial brands like Ultratech onto their floors to clean them. But little systematic study of the natural roles of oil-eating microbes had occurred until Deepwater Horizon, when the field became a full-fledged science.

Joye was a short, serious, eyeglass-wearing thinker who had studied the topic long before anyone ever took an interest in it. Growing up on a North Carolina tobacco farm, a walk-on who made the University of North Carolina basketball team, in school she was a defender of the defenseless. She became entranced by the Gulf of Mexico on her first deep-sea submersible dive and went on to a distinguished career at the University of Georgia. The day Deepwater Horizon exploded, she was on the Gulf doing research. She immediately turned the boat to the site.

Petroleum hydrocarbons form natural seeps in the Gulf of Mexico and elsewhere in the world. Microbes such as *Colwellia*—named for the microbiologist Rita Colwell, who taught Indian women to prevent cholera by wringing drinking water through their saris—became hot areas of study as a team of researchers struggled to understand how to devour some fifty million barrels of oil. "I love microbes that eat petroleum!" Joye announced as the keynote speaker at the New Orleans meeting of the American Society for Microbiology in May–June 2015. Only a few years earlier, the organization had rejected her offer to speak.

Joye came under the media's bright spotlight when she told the world much more oil was spilling out of the discharge than BP and the U.S. government claimed. Funded by the National Science Foundation, the Environmental Protection Agency, and the Gulf of Mexico Research Institute, she directed a team to analyze and advise the efforts to contain the rapidly expanding disaster. She triggered headlines by insisting that 75 percent of the spilled oil was still present. Short, blonde, and often wearing a green sweater, she reminded the packed newsrooms, "I have a loud voice and a southern accent!" as she described the beauty of the Gulf and of its rich living community of organisms that thrive in oil. A natural petroleum seep will feature magnificent sea worms, hydrate mounds, and numerous microbes grazing on methane and ice, she lectured. There are some 22,000 such natural Gulf seeps, of which 1,000 were significant enough to be visible from satellites. The best estimates are that nature itself leaks 2,500 barrels of oil per day.

The emergency workers had cut the riser pipe in the first hours after the blast. The platform collapsed, creating several release points of gushing oil. The cut was completed on June 3, but the first attempt to seal it failed. Two colors of the gushing fluid could be seen—oil, and oil-plus-gas, which was darker in color. June 6, 2010, represented the maximal outflow, Joye said, "an astounding and disheartening 29,000 barrels a day." The total, according to government experts when a permanent cap was placed, was somewhere between

4,900,000 and 6,000,000 barrels of oil and gas. The figure may have been higher, and the well continued to leak.

What she saw underwater were dozens of jetlike deep-water plumes marking a "sudden microbial feeding frenzy" formerly regulated by the scarcity of nutrients. *Colwellia*, aromatic-eating *Cycloclasticus*, alkane-eating *Oceanospirillales*, oil-eating *Alcanivorax*, *Neptunibacter*, and methane-loving *Methylococcacea* dominated, as the usual natural diversity went missing. Microbes that loved to eat dispersants "had a delicious feast in what amounted to a six-week experiment," Joye said. Organisms that eat sugars also thrived, such as *Colwellia* and other Proteobacteria. Further analyses published in *Nature Microbiology* showed more species were capable of cleaning the oil than previously thought.

At the same time an international team of American, South African, and Dutch researchers was descending rapidly down the deepest mine ever built, 4.2 kilometers, to the 155-degree heart of the Mponeng gold mine. "I've never seen a face get so red," the guide warned the fifty-three-year old Princeton geologist Tullis Onstott. He slumped and turned the lead over to the University of Freestate biochemist Esta van Heerden, another woman who, like Joye, loved extreme microbes, in her case those that eat radioactive uranium, chromate, and other heavy metals in the world's deepest mines. The thriving, fascinating, diverse, living ecosystem in mines like Mponeng suggested that microbes could potentially help restore radioactive wastelands.

The hardiest microbes of Mponeng included *Clostridium*, relatives of Chaim Weizmann's acetone producers. Members of the *Clostridium* genus live in the human gut as microbes that can cause infection but also, paradoxically, dampen inflammation. *Clostridium botulinum*, which produces one of the strongest toxins known to man, makes the key ingredient of Botox injections. The deep-mine, radioactive-friendly microbes renewed the interest in potential Mars life.

In short, the conditions for microbes were present almost everywhere on Earth—not only inside of us, but also in the harshest environments imaginable. They could live in space. They use almost every known energy source in the universe. They clean industrial, water, radioactive, and petroleum pollution. The problem was, they were slow.

"This is a golden age of exploration," the dark-haired Onstott said to me from his basement office at Princeton, describing his work in South Africa and deep below the surface of the Canadian Arctic, where he assisted NASA in determining how to detect microbes elsewhere in the solar system. "There is a ton of microbial dark matter down there," he said of the DNA they extracted, "a shadow biosphere" that rivaled the visible biosphere in mass. Because of his expertise, Onstott was a key NASA adviser on the Mars missions.

Then a new comet, discovered by a photographer and an astronomer in Russia, became the object of the most unlikely search-for-life mission of all.

Rock star

In December 2014, after a ten-year, five-billion-mile journey, the tiny European Space Agency probe Rosetta was nearing its target—a barbell-shaped comet called Churyumov-Gerasimenko, somewhere beyond Jupiter's orbit. The dramatic landing of the three-pronged lander from Rosetta named Philae, spinning in space as it left the mother ship and hurtled to the craggy surface, captivated the whole world as it was broadcast on the web and television. The audience hunched over computers as the tiny wounded lander dove to the surface 320 million miles from Earth. It was a mission so fraught that NASA had turned it down as impossible.

After ten years in space, Philae landed late in 2014. It made a media sensation of a magnitude not seen since the moon landing. Television, Internet, and live Twitter feeds from Philae had people posting nonstop from all over the world about being inspired to be human. I followed the congratulatory speeches and media reports on my laptop, much as I had watched Neil Armstrong on black-and-white network television in 1969. The coverage came complete with a re-searcher's Bahama shirt controversy featuring naked women and millions of fellow watchers in Europe, Asia, and the Americas. Even the mayor of the tiny German town that was Rosetta's command center was teary-eyed.

The lander fired in, however, much too quickly. The team knew that only two of its three bayonet-tipped legs were working. In a moment of long-awaited, brief triumph, however, the team screamed with joy when it made it to the surface and sent mankind's most stunning pictures to date of the solar system's beginning—in the form of an anvil-shaped, four-kilometer-long rock.

But the Philae lander was dangling on the comet's dark side, in a shadow, sampling rocks and radioing back readings. The team debated whether to fire the rockets to keep it upright or save the energy and do more science. It opted for the science, which paid high dividends when the ESA team found unusual heavy water on Churyumov-Gerasimenko, not like Earth's, calling into question once again the origin of Earth's water. Philae lasted for twenty-one and a half hours and then died. In June 2015 Philae awoke and shared some data as the comet spun briefly into the sun's view. It then fell back to sleep, with the only hope for recovery shortly as it would dive into the sunlight.

Rosetta revealed thirteen new amino acids not found on Earth. It detected a porous interior, matching a lab simulation revealing that comets resemble fried ice cream in dirty snowballs, crusty on the outside and porous on the inside. It also pictured sand dunes, hardly anticipated on a comet surface. As the comet raced to the sun, Rosetta analyzed particles blasted off the comet's surface, detecting organic carbon compounds in two particles team members named Kenneth and Juliette. With Philae perched in the shadow of a cliff, scientists debated whether to expend its last energy to drill, hoping that would raise it into a more upright position. It did not work, but it did reveal the water on Churyumov-Gerasimenko to be isotopically distinct from the water on Earth.

We were going through a revolution in our imaging and high-resolution analysis of the beginning of the solar system. Meteors and comets, we knew, contained at least eighty amino acids. There were sixty thousand asteroids and 157 known comets. In the lab some had sown life in the soils analogous to that of Mars, and many think its building blocks were transported from there or comets. "It's a golden age," said NASA's Michael Mumma to me after a conference session, "for the discovery of life's building blocks in space."

Meteors, dust clouds, the ice on Saturn's moons Enceladus and Titan—analysis of the universe's debris proved it chock-full of hydrocarbons and amino acids. The unfolding series of discoveries reenergized the quest for life's origins. At the same time, on Earth a nascent bioremediation industry turned to microbes to clean wastewater and other toxic spills. Acceptance of microbes' critical roles in health and the environment seemed right around the corner.

The idea spread that relatives of those same microorganisms that lived in the depths of the earth or in hot vents below the sea also lived in our intestinal tracts and those of farm animals. Suddenly the insights from microbiological cooperation became a target of medical funding as global epidemics in antibiotic-resistant bacteria gained attention.

At the NASA conference in Chicago, Sandra Pizzarello quizzed me on my book.

"Here, you finish it." She handed me half of her salad. She stood up to head to a conference session. and I looked out at a gathering rainstorm on the street.

Out on Michigan Avenue, in the June summer heat that built anvil-shaped clouds over the lake, car horns blared and the sidewalk shimmered. "Tell me if you need anything more," Pizzarello called, as the city swept by me and pedestrians rushed before the storm. Hunched and running to the El, I thought of Saul Bellow's character Tommy Wilhelm in *Seize the Day*, finding himself in the "great, great crowd, the inexhaustible current of millions of every race and kind pouring out, pressing round, of every age, of every genius, possessors

of every human secret, antique and future, in every face the refine-
ment of one particular motive or essence—*I labor, I spend, I strive, I
design, I love, I cling, I uphold, I give way. . . ."* I felt, suddenly, alive to
the strivings of many organisms, above and below and inside of me,
and realized how not alone I was.

A wider acceptance of the microbes' critical roles, in human health
and the global environment, seemed right around the corner. Then
some of the claims fell apart.

PART TWO

Turning the Tide

Seeking Better Health

6

Killer Membranes

The Labs Where Life Is Made

The Szostak Lab stood in Harvard's sleek Simches Building, filled with top researchers from around the world. The glass-lined doorways opened into a cool, temperature-controlled environment with state-of-the-art genetic equipment flanking the tomes of nineteenth-century discoverers. An Israeli-born doctoral candidate named Itay Budin peered anxiously to see if the coveted chemicals had come alive.

Growing up in a peripatetic household whose residences ranged from Germany to Montreal, the Harvard researcher Jack Szostak loved reading the biographies of great scientists. In the 1990s the Canadian Szostak helped shape the molecular-biological revolution with the Nobel Prize–winning discovery, with his collaborator Elizabeth Blackburn and her associate Carol Greider, of an enzyme that preserves chromosome endings and could possibly extend human longevity. The discovery shocked the world and ignited a business and science race for a longevity drug. But Szostak was not interested in longevity. He wanted to make life.

The search for the origin of life had been a roller coaster since the days of Miller and Urey and the exploration by the Salk Institute's Leslie Orgel and others of conditions under which nucleic acids could replicate. In 2003 Miller had despaired of making more progress, but regained his optimism later, at the end of his life. The discovery of amino acids on meteorites helped revive the field, setting the stage for a dramatic International Society for the Study of the Origin of Life (ISSOL) meeting in Oaxaca, Mexico, during which the organizer, David Deamer, related the origin of life to star death and planet birth and the interfaces among minerals, water, and the atmosphere on Earth. But nobody was going to see a star or planet birth, at least not up close, and certainly not Earth's.

One alternative was to create life in the lab. Since 2000 that was what researchers around the world had been seeking to do. Some tried genetics first, by manufacturing a simplified version of RNA. Others tried metabolism first, by devising simple experiments that mimicked modern biochemical pathways. Some worked from the bottom up, others from the top down. In Boston, Szostak combined those approaches, invoking chemistry, molecular biology, and genetics in the quest to solve the problem, funded by NASA, the Department of Energy, the Howard Hughes Medical Institute and other private investors. Competing governments supported other researchers.

By the late 2000s Szostak, with the graduate students Irene Chen, Itay Budin, and others, had discovered that synthetic cells in the lab could be made to evolve into more and more efficient, viable, and stable versions of themselves. With fatty acid walls made of compounds that might have formed on meteorites, the "cells" could seemingly grow, compete, and reproduce. But the idea of making life from scratch, or engineering life to do what we need, was hamstrung until a series of remarkable discoveries in several labs. Szostak sought to build a cell in the genetics-first mold, while others, like the Rome researcher Pier Luigi Luisi, believed that cell membranes came first. In Tokyo researchers sought to combine the two, while in Maryland and California the genome researcher Craig Venter sought to find the minimum requirements for microbial life and went to find its true diversity on the high seas. Together they revolutionized our ideas of what life is and what scientists could do.

For Venter, synthetic or designer microorganisms opened a new industry. In 2005 he founded the company Synthetic Genomics to engineer new microorganisms that could potentially produce alternative fuels, deliver medicines, and even generate energy. In hospitals and school locker rooms, for instance, antibiotic-resistant infections were creating a crisis. Young people were becoming violently ill from multidrug-resistant *Staphylococcus aureus* and methicillin-resistant *Staphylococcus aureus* (MRSA). Some proposed engineering microbes to deliver better therapies directly or indirectly as carriers of other medicines. A race was on for the fourth kingdom of life, dubbed the "synthetic kingdom." "Are you playing God?" an NPR reporter once asked Venter.

"We're not playing," he replied.

As interest deepened, promising and arcane lines of pure science became the focus of intense medical and industrial competition. There was little oversight, and some companies marketed microbes for people with emotional problems while others claimed to treat diabetes and obesity. As the money increased, the task became one of separating the truth from spurious claims. One of the first people there, with a pure-science desire to uncover life's origin, was Jack Szostak.

Synthesizing life

The origins of synthetic biology date probably to 1828, when the German chemist Friedrich Wöhler created urea, the organic compound in urine, from inorganic ammonium sulfate. "I can make urea without the use of kidneys," he wrote to the chemist Jons Jacob Berzelius. Wöhler admitted that he was depressed by the fact that, in so doing, he had disproved a central tenet of vitalism, the idea that organic compounds are somehow divinely different from inorganic chemicals. "The great tragedy of science," he purportedly added, "the slaying of a beautiful hypothesis by an ugly fact." Years later, however, urea would be a key component of microbe-powered fuel cells and Wöhler's tragedy a foundation of organic chemistry and astrobiology.

There was nothing new about the dream of remaking life to do our bidding. All of the farmers in all the eight-thousand-year history of human agriculture, animal husbandry, fermenting, and gardening had manipulated biology. So too had the practitioners of the nineteenth-century dream of reanimating dead tissue that gave us numerous traveling charlatans and Mary Shelley's *Frankenstein*. In 1911 a University of Durham botany professor, M. C. Potter, described generating electricity with microorganisms such as *Saccharomyces* or bacteria engaged in "fermentation and putrification." The molecular-biological revolution of the 1980s and '90s produced the human genome, the transplantation of genes for lab study, and claims of rational drug design. By the 2000s some of the same people who had led the race to sequence the human genome began gravitating to the strange, enticing field of creating or manipulating microbes to do our bidding.

Born in 1952 in Montreal, the year Miller and Urey made their discovery, Jack Szostak found his passion in grade-school math class and in the chemistry lab his father built for him in the family's suburban basement. Szostak's Polish-born parents let him play with dangerous chemicals, and he blew up one concoction so emphatically that it left a piece of glass embedded in the ceiling. Around that time the journal *Chemical and Engineering News* called for an American national goal of designing cells to meet our energy and environmental crises.

Szostak graduated from McGill University at the age of nineteen. While visiting Cambridge, England, Szostak explored the Laboratory of Molecular Biology and met the future Nobel Prize winner Sydney

Brenner, whose project using the worm *Caenorhabditis elegans* for developmental genetics ignited his imagination as to the possibilities of molecular biology. Szostak then went on to complete his Ph.D. in biochemistry at Cornell University and afterward went on to Harvard Medical School to establish his own lab at the Sydney Farber Cancer Institute. In 1988, at the relatively youthful age of thirty-six, Szostak was granted tenure at the Harvard Medical School.

In the early stages of Szostak's career, his team focused on creating the world's first artificial yeast chromosome. At the Sidney Farber Cancer Institute in California, he cloned yeast telomeres in collaboration with the labs of Elizabeth Blackburn and Carol Greider. Telomeres are the tips of chromosomes; they fray as we age, much like shoelace ends. The team discovered a chromosome-lengthening enzyme that could be a drug target for longevity, helping to ignite a science and business frenzy. In 2009 it led to a Nobel Prize for Szostak and his colleagues.

Already Szostak was pursuing fundamental questions relating to the origin of life. Cech and Altman's discovery of RNA enzymes had propelled him away from genetics and down a different path entirely. "[Cech and Altman] inspired me to try to think of ways to make RNAs in the lab that could catalyze their own replication," Szostak told *Nature*. In 2001 he, Rome's Pier Luigi Luisi, and San Diego's David Bartel published the short paper "Synthesizing Life," discussing the membrane biophysics of prototype cell walls. The goal was to understand how a set of molecules first came alive. "We assume we have the chemical building blocks of life," the three wrote. "The question we're looking at is what we need to do to make these chemicals work like a cell?"

Szostak's lab and others competed to make a protocell, the precursor to modern cells. Protocells had to be much simpler than modern cells. They had no complex proteins, presumably; rather, they consisted of molecules gathered inside a fatty membrane that somehow became capable of replication. The first protocells copied themselves with slight variations, some variations having more advantages over others, leading to more successful protocells. That was the idea, at least.

The Szostak lab began by creating fatty-acid prototypes of cell membranes. Then they injected RNA into these primitive bubbles and showed that genetic material promoted cell growth at the expense of other protocells that lacked RNA. With a bit of shaking, as in a tidal pool, the protocells lengthened into tubes, with hair-like tails, that broke up into two daughter cells—a primitive form of reproduction. In 2004, in a shocking discovery, Irene Chen, a member of Szostak's lab, injected fat into the synthetic cells and watched them rapidly develop tubes and hairs. "I had joined the lab because I thought the experiments were amazing," she told me. "I was willing to work on anything cool." Itay Budin showed later that the more complex the cell's contents, the better it competed. By 2013 the lab had created a prototype of an artificial cell. But what they needed was the capability to replicate RNA molecules without enzymes, a difficult problem because RNA easily fell apart in water. They needed to bring more energy to the system.

To bring energy, Itay Budin turned to the extreme temperature gradients at volcanic vents. With a friend he went "junkyard diving" for hot and cold plates in an abandoned neighboring lab. In a cold room with a giant pair of pumps pushing neon-blue antifreeze for the icy ocean, and a hot plate to mimic a volcano, Budin waited. By setting up a capillary tube between the two, they showed how big temperature fluxes induced the self-assembly of membranes and even the co-assembly of cell-like vesicles containing DNA. "It was really a home-brewed experiment," he told me, sitting in his Berkeley office. "I was a little surprised at how well it worked."

Dozens of researchers now competed to make a protocell, or minimal cell, or artificial cell, from top down or bottom up, including Tetsuya Yomo at Osaka University, Yusuke Maeda at Kyoto University, Vincent Noireaux at the Rockefeller Institute, Jerry Joyce and Dave Deamer in California, and NASA's Winona Vercoutere. The quest regained momentum with grants from the U.S. businessman Harry Lonsdale, the John Templeton Foundation, the Simons Collaboration, and the Harvard Origins of Life Initiative, and money from Craig Venter at his well-funded institutes in Maryland and La Jolla. One of the key insights came from the Italian Pier Luigi Luisi.

In Tibet

Born in 1938, the biologist Pier Luigi Luisi was a pioneer of the study of the origin of life. His Rome lab was studying the origin of life by building cell walls first. He reasoned that before you could have genetics, you needed a place to hold the genetics. Practicing science in an interdisciplinary way, hosting an annual conference on biology and the arts, Luisi was an active author who saw that there must have been a bottleneck early on between competing processes and networks that led to the current dogma that DNA makes RNA and RNA makes protein. What was the bottleneck, and why did it get solved in just this way? he wondered.

In 2009 Luisi organized an origin-of-life conference in Rome to discuss the problem. You cannot create self-replication with a single molecule, he told the group. You need enough material to build an active binary complex so that one molecule can replicate, so you need to make this self-replicating RNA. But no one could. "Until somebody finds a way by which RNA replicates, the RNA world is baseless," Luisi said. Others around the world sought to find exactly that.

In 2011 Luisi found that key cell functions can be instigated with just eighty genes, but he wanted to get the number down to fifteen or so. He was invited to speak to the Bhutan Institute of Mind and Life in a remote mountain monastery. With the Dalai Lama as director, the Institute sought to link Western science and Buddhism. Luisi flew from Bangkok to Paro, where he was met by—nobody. He finally managed to reach the Sishina monastery, home of ninety Tibetan Buddhist nuns ranging from thirteen to sixty years old. The nuns knew the DNA double helix, they followed DNA replication, and they understood genes and how they encode for proteins and cause diseases.

But they asked an unusual question: "Why does DNA replicate? How does it decide to do so?"

It was not that the gene decided to replicate, he said. The whole organism gave the message, created through a complex cascade of ions, hormones, and other chemical activators, by which the organism message "arrived at the DNA in the form of . . . molecules." The nuns followed his antireductionist, systems-biology speech quite well. They called it "cooperative arising." They took it as an example of Buddhist impermanence: anything born is destined to die. Perhaps that was the definition of life: something capable of dying.

The Tibetan nuns brought Luisi a new understanding of the battle to create life. The origin of life was biology's most interdisciplinary organizational problem of our time. The scientists were all fighting about the sequence of what came first in a microbe. Was it genetics, metabolism, or something else entirely? There was a protocell faction, people like Szostak. The minimal-cell faction sought the smallest number of genes to make life, like Luisi. But an upstart group, the artificial-cell faction, wanted to try from the top down, in the lab. From this group a familiar name ignited the field.

The speed of light

In the late 2000s the genome researcher Craig Venter had seemingly done everything he wanted in science and made a fortune and a few enemies while doing so. Venter's drive for the full human genome sequence outcompeted a government group and shared credit with them in 2001. Since then he had created and lost two companies. He divorced his second wife, the biologist Clare Fraser, in 2005 and subsequently married his publicist, Heather Kowalski. The former CEO, owner of yachts, and breeder of dogs had everything one could desire.

But what he desired wanted most was to create synthetic microbes to address the crises facing the world—energy, waste remediation, global warming, bioterror. As he said in 2010, "Designing and building synthetic cells will be the basis of a new industrial revolution. The goal is to replace the entire petrochemical industry." One could program cells to do what we needed. He was in an all-out effort to overturn yet another field.

Growing up in southern California, Venter had struggled in school and preferred surfing and sailing to science. Drafted to fight in Vietnam, he placed second on an exam as an Army medic and saw the unimaginably fragile boundary between life and death. He returned to community college and, eventually, to a doctorate in neurobiology. After a rocky stint in neuroscience at the National Institutes of Health, he partnered with a billionaire investor to found the Institute for Genomic Research (TIGR), which raced the government to compile the full sequence of the human genome. He gave his community-college composition teacher a job as the director of communications. Balding at seventy-four, he still had the athletic powerful shoulders and arms of a competitive swimmer. He had teamed with an ex-Navy engineer, Hamilton Smith, the ballast to his driving personality.

Together they hatched a plan to sail the seas looking for new microbial life. In 2003 and 2009 Venter led two ocean expeditions, collecting water samples every two hundred miles to document new microorganisms' DNA. The first voyage touched areas of North America, Australia, and Antarctica. The second sailed in 2009 and touched areas of unique biodiversity around the shores of Central America, southern Europe, and other locations. Venter sequenced the new genomes, claiming to discover the incredible number of eighty million new genes in the process. The trouble was, no one knew what those genes did.

In the lab, he applied for and won a Department of Energy grant in its new "Genomes to Life" program seeking ideas for harnessing microbes to create new energy sources. He told a TED audience in February 2005, after founding his company, that "we're moving from the stage of reading the genetic code to actively writing it" and devoted his lab to doing so. In 2008, after he announced they had created a bacterium with engineered, artificial DNA by adding foreign genes to the original genome of a bacterium, they synthesized an entire genome. The offspring also carried the artificial DNA, showing that a manmade microbe could be tailored to specific purposes, such as producing antibiotics, cleaning up pollution, or increasing crop yield.

People had been adding exogenous genes for ten years, and the quest for synthetic life as an electricity source went back eighty years. In 1931 the researcher Barnet Cohen had created primitive microbial fuel cells. In the 1990s the engineer Oliver Peoples had worked with his company Metabolix to develop microbe-made degradable plastic utensils and bags. DuPont put out a new, springier type of carpet made in part from fermented corn, which required 30 percent less energy to make and reduced greenhouse gas emissions by 63 percent over normal carpet manufacturing. The University of Wisconsin's Fred Blattner sold a pared-down "clean genome" of *E. coli* to be made to do whatever a customer wanted—make hemoglobin, for instance. At Harvard, Pamela Silver engineered gut microbes to report on the health of the host body and worked with Dan Nocera on using bacteria to create a solar-powered "bionic leaf." "Any biological system is a kludge," she explained to me. "It makes itself good enough to get by. But to replicate it in a manmade system is very hard."

On the electrical front, Penn State's Bruce Logan, supported by the National Science Foundation (NSF), was developing microbe fuel cells that could generate electricity while treating wastewater. They proved they could produce electricity from ordinary domestic wastewater as well as animal, farm, food-processing, and industrial wastewater and hoped to demonstrate it at larger scales. Several people, including the South Korean researcher B. H. Kim, created microbe-powered batteries. A Clinton, Iowa, plant was creating plastics made by corn-fed microbes, while other microbes could detect arsenic in drinking water.

In 2010 Smith and Venter engineered a synthetic-biology breakthrough, creating in the Craig Venter Institute a completely artificial microbe. They replaced the entire genome of a living one-celled organism with a manmade sequence of synthesized DNA. The procedure involved hundreds of thousands of failed attempts, including a replacement of all the equipment and a switch from a bacterium to a fungus as the target at one point. Their success meant he was hastily called to Washington to meet with the director of the National Institutes of Health, Bernadine Healy, to discuss the prospect of withholding publication to prevent bioterrorists from hacking the technique. Never one to opt for either humility or patience, Venter wrote that his Institute "immediately set to using the techniques to design microbes for toxic waste cleanup and energy production." In July they created the first synthetic cell, and in May 2013 a synthetic flu vaccine. By May 2014 the company had automated DNA, RNA, and protein production, and in September 2014 it engineered a bacterial virus called a phage. In March 2016 it created the first minimal artificial cell.

Around the world, money for the synthetic life movement poured in from private investors and Kickstarter campaigns. Harvard's Pamela Silver tapped students, cofounding a global student competition sponsored by MIT. The genetically modified (iGEM) student award was meant to spur the development of new ideas, resulting in proposals such as one from the Netherlands to use 3-D printers to create new microbes, from the Czech Republic to use cellular communication to identify tumors, and from Israel to use the heralded CRISPR gene-editing technology to target cancer cells or pathogenic cells and to reengineer human microbes to improve metabolism and even sun protection. The 2016 competition featured high school, college, and graduate teams from five continents, funded by eight companies, with the winning undergraduate team from Imperial College London producing a technique for cofounding organisms impossible to grow alone. The annual conference in Boston developed its own subculture, with students from all over the world often dressing in costumes, offering ideas that often were more creative and advanced than those of adult science.

A new industry, dubbed Synbio (for "synthetic biology"), sought to commercialize the discoveries with some of the whimsy of a new biology-hacking subculture. European companies, fueled by investment from their governments, led the way, followed closely by those the countries of the Asian Pacific Basin. In the United States companies such as SERES, Synthetic Biologics, Syngenta, SynbiCITE, BioRealize, and the One Sky Initiative shopped for investors at business conferences. The startup MapMyGut made do-it-yourself kits for identifying the constituents of your one's microbiome. Sapphire Energy of San Diego engineered algae to produce oil, aiming for five thousand barrels a day from its New Mexico algae farm in 2018. The BioBricks Foundation created an AddGene website to share artificial gene components, including an entire cloning toolkit from England's Sainsbury Laboratory, for free. A startup called Growing Plant sought to create luminescent trees for street lighting from naturally luminescent bacteria and houseplants for reading lamps. San Francisco's Twist Biosciences produced synthetic DNA for other labs to commercialize.

The company Synlogic developed a cancer-detecting microbe, and in 2016 the Chicago pharmaceutical company AbbVie, maker of the world's top-selling drug, purchased a Cambridge, Massachusetts start-up with the idea of programming a bacterium to sense the waxing and waning of an inflammatory condition and to activate a drug accordingly. Evelo Therapeutics, Ernest Pharmaceuticals, and Actogenix all sought to design therapeutic microbes.

As an economic model, synthetic microbes faced huge obstacles in moving from early-stage financing to marketable products. The regulatory environment was murky. The U.S. Food and Drug Administration sought to keep up with the science. Microbial fuel cells were weak, limited by the low surface-to-volume ratio, since electrons can be generated only when bacteria contact the electrodes. Petroleum-waste remediation by microbes was slow. The technologies were improving rapidly as business investment and media attention increased. The amazing thing was that the metabolisms that drove a cutting-edge industry were hundreds of millions or billions of years old.

"A world of difference"

Imagine the Earth four billion years ago. Volcanoes spew poisonous gas over a toxic ocean dotted with only a few tiny islands resembling Iceland or Hawaii. The equator is possibly broiling hot and certainly dangerously irradiated, the poles bitterly cold. The sun is smaller. The oceans have a completely different chemistry from today's, more acidic and hot—what some say global warming might portend. Shallow ponds on the tiny islands fill with rain, dry out, and fill again. Comets rain down on Earth on the order of dozens or more each day. If a cell lived, it would be cooked, all water evaporated and its proteins dry-roasted to charcoal. Ice ages may have intervened. Somewhere in one of those ponds, hot as Death Valley or in temperatures well below zero, life began. Szostak argued that this natural world was completely different from the Venter's technical wizardry. "Creating synthetic genomes," he wrote, "may ultimately let us redesign existing forms in ways we can scarcely imagine [but] there is a world of difference between artificial life and the recent feats of Venter and Smith."

What Szostak wanted was to get template-directed RNA or DNA polymerization to work. He was discovering that in RNA, it did not work. He changed one carbon-oxygen bond to a carbon-nitrogen bond. He moved the nitrogen from the 3 position to the 2 position. Every time he tried to solve a problem, he made the premise a bit more unnatural. "He'll end up with a nucleic acid alphabet with twelve letters instead of four, but he's going to get there kicking and screaming," said NASA's Steven Benner.

In Florida, Benner was focusing on the compound borate as the template for the source of life. The advantage of borate, he argued, was that when coupled with another mineral, it prevented organic compounds from simply ending up merely as "tar." In another approach, John Chaput, then at the Biological Design Institute at Arizona State University (ASU), was trying to create genetic systems by modifying the sugar backbone. At ASU's Beyond Center, Sara Walker saw life's beginning as a process, not as a material, and that process involved the nonlinear movement of information in a biological system. She told the *Huffington Post* she was "incredibly optimistic" about uncovering life's origin. Piet Herdewijn, a Belgian researcher at the Catholic University of Leuven, worked with Philippe Marliere. Ichiro Hirao in Japan was at the cutting edge of the Riken Institute, and the Japanese Research Institute was working on genetic systems.

So far, synthetic biology had not lived up to its hype. There was still no conceivable manner by which to imagine inanimate chemical compounds suddenly becoming alive. The sum of Venter's series of experiments was to show how little we knew of the operating system of the living cell. Szostak's forays had involved so many substitutions as to become, arguably, as much a string of brilliant lab exercises as a viable exploration of nature as it really worked on early Earth. The same could be said of virtually any synthetic-biology foray. Perhaps it was for the best not to resurrect spontaneous generation or Frankenstein just yet, because the claims for microbes in mental and emotional health were fast outstripping the results. "People are obviously desperate for solutions," the Texas biologist Mark Lyte told the *New York Times* in June 2015. "It's the Wild West out there."

Scientists were routinely making synthesized sugars in the lab. In 2013 the Szostak team got RNA to replicate in the lab. It meant that synthetic life had acted like real life. In 2014 Jack Szostak predicted that artificial life "could be achieved in the next few years," but then he backtracked. The idea caught on as a series of crises brought the potential markets for synthetic biology to the world.

Hold the medicine

By 2012 antibiotic resistance was becoming the "number one health crisis in the globe," said the director of the World Health Organization, Margaret Chan. High school athletes were dying from infections of the antibiotic-resistant microbe called methicillin-resistant *Staphylococcus aureus* (MRSA). The puppeteer Jim Henson died of an antibiotic-resistant infection. In hospitals, nursing homes, and medical clinics, by 2012 the battle against antibiotic resistance was reaching a crisis. The American Medical Association called it "the biggest health challenge of the 21st century."

At the same time, the use of probiotics for tailoring microbes to a patient's needs became an acceptable treatment for autoimmune conditions like Crohn's disease and irritable bowel syndrome. Other research focused on the role of microbe loss in rapidly rising rates of obesity and diabetes. For seventy years American agricultural companies had fed antibiotics to cattle to fatten them more quickly and with less food, thus killing some of the gut microbes that kept them slender while also avoiding some illnesses. Much the same seemed to be happening with children.

In the United States, the director of New York University's Langone Medical Center, Martin Blaser, explored the links between the microbiome and childhood obesity, which had gone up some 450 percent in fifteen years, and the rise in antibiotics in humans and in animal feed. Laurie Cox, in research published in *Cell*, discovered that if you limit antibiotic exposure for four weeks at the beginning of life, gene expression in mice intestines is improved and weight gain in later life is reduced. Giving antibiotics also fattened sterile mice. The penicillin caused inflammation in the gut, but it was not clear whether the diminished number of microbes played a role. NYU's Blaser showed that the states in America with the highest rates of antibiotic prescription also featured the highest rates of obesity. But not all mice or humans who received antibiotics in infancy became overweight, so what were the differences between those who did and those who did not? That question became the focus of several studies.

The Washington University researcher Tanya Yatsunenko compared the gut microbiomes of U.S. babies with those of Malawians in Africa. At three years old, she and others discovered, the transmission of microbes from mother to baby helped in developing better immunity, metabolism, and cognition. The womb itself is sterile, but when the water breaks, a mother's bacteria ascend the birth canal. During birth, as the newborn moves through the birth canal, it picks up more of the mother's microbes. A baby born by cesarean section can be swabbed with fluid from the vagina with much the same effect, but also with the danger of infection. In the child's first three years it picks up helpful microbes through kissing, licking, premasticating food, and the mother's nipples in breastfeeding. In modern countries, where bottle feeding was prevalent, some of these modes of transfer were being lost.

As for probiotics and mental health, desperate parents whose children had not responded to conventional drugs drove the science, creating a dangerous temptation to overhype claims. Iowa State's Mark Lyte proposed that probiotic bacteria could be tailored to treat specific psychological diseases. Lyte had been pursuing the idea since the 1990s, but only as technology improved did it take hold. There was evidence of increased risk for schizophrenia associated with prenatal exposure to influenza. Some children developed obsessive compulsive disorder (OCD) or tic disorders following a strep infection. Despite continuing skepticism, evidence mounted in support of pediatric autoimmune neuropsychiatric disorders associated with streptococcal infections (PANDAS). What was needed was proof.

In 2006 scientists at Columbia University asserted that up to one-fifth of all schizophrenia cases were caused by prenatal infections. Columbia University researchers also demonstrated that strep-triggered antibodies alone are necessary and sufficient to trigger a PANDAS-like syndrome in mice. In Ireland, University College Cork scientists showed that mice that ingested the *Lactobacillus* bacterium experienced better emotional health, via the vagus nerve connecting the stomach to the brain. A Canadian team showed that antibiotics could be linked to risk-taking behavior in mice. A microbiologist at Stockholm's Karolinska Institute showed that mice raised without microbes spent more time running around outside than healthy mice in a control group, with more recklessness. A study in 2016 showed that fecal transplants improved the gastrointestinal health and behavior of some eighteen autistic children. "By changing what's going on inside of the gut, we hope we can change how the brain responds to the environment," said the study head, Kirsten Tillisch, of the University of California, Los Angeles, of her research, funded by Dannon Yogurt.

Companies seized on the science to promote probiotics for the treatment of weight and neurological disorders. They included, among others, Flora and Gut Flora, Ecologic Barrier, and Digestive Advantage and Probiotics Solutions in the United States, Yakult and Morinaga in Japan, Winclove Probiotics and TNA in Holland, Vizera in Slovenia, Shenzhen and Shanghai in China, Symbiofarm and Organobalance in Germany, Sofar and Mofin in Italy, Protexin in the United Kingdom, Pileje and Ninafarm in France, Probi in Sweden, and many others. *Lactobacillus* emerged as an early hero, showing that healthy volunteers who received it reported significantly lower stress levels than those who received a placebo. Several researchers explored the ways gut microbes influenced the levels of neurotransmitters such as serotonin. The global synthetic biology market, of companies seeking to redesign microbes for medicinal and industrial use, was projected to range from $11 billion to an astounding $37 billion by 2022. The wave of investment, thought, and promise rose to meet the challenge.

For all the business interest, however, the lab quest for synthetic life hit yet another impasse. Despite the improvements in technology and the best efforts of well-funded labs, the fact that water was almost universally accepted as necessary for the onset of life created a stumbling block. RNA falls apart in water. The same was true of the sugars and phosphate compounds that make up the backbone of RNA and DNA. On Earth they are generally available only in a solid, stable state in the environment and are thus unavailable to protocells. The suspicion arose that maybe our whole effort was off track. "After a point," said NASA's Steven Benner, "if you're failing over and over, maybe there is something wrong with the entire approach. Maybe we need to start again."

The essential question for the origin of life was how matter could transition from a nonliving to a living state. The answer was critical to finding microbial life on other planets. Most studies had focused on DNA, protein, lipids or cell walls, or metabolic processes. A new approach was necessary, a growing number of researchers felt, one that focused on the processes of information flow in physical entities.

Then an insight came from an unexpected angle.

Relics of the Deep

Future from the Past

The geologist Andrew Knoll was fifteen years old when he hiked up amid the oaks towering over the rich dirt of the Marcellus Shale formation near his hometown of Wernersville, Pennsylvania. He explored the foothills of the Appalachians, finding fossils of tiny Paleozoic creatures of the sea floor, like brachiopods or trilobites. At Lehigh University, while giving a report on Lynn Margulis, Knoll became convinced that life and Earth coevolved. At Harvard University he traveled the world reading the testimony of rocks, seeking evidence of ancient life in their wrinkles and inspiring students like Nora Noffke to understand the microbial dance with global geology and climate change.

The deep relationship between geology and microbial life runs throughout Earth's history. Without one, you could not have the other, not in the states they exist today. Simulations suggested strongly that Earth's mineral makeup is probably unique in the universe as shaped by its microbes. Microbes altered much of the world's structure, creating Dover's White Cliffs, Australia's banded iron formations, and many of the minerals from which humans made our first tools. They likely seeded the clouds for rain and snow, certainly devoured petroleum spills, helped to solidify and deposit radioactive and mining contamination, and oxidized the manganese and iron at ocean ridges and in backyard soils. In South Africa, Siberia, China, and northern Australia, Knoll found some of the earliest known fossils of microbes. He advised NASA's Mars missions. He and the geologist Robert Hazen, at the Carnegie Institute, suggested that the origin and processes of early life were intimately connected with the origins and characteristics of many minerals.

As Knoll travelled to the islands of Norway, to the Siberia River, to Namibia, and to Australia, Hazen did the same in South Africa and Asia. At the Carnegie Institute Hazen was recreating the steps by which life might have started at deep-sea volcanoes that house teeming communities of giant tubeworms, lobsters, and sulfur-loving microbes. In his lab Hazen recreated the vents' high heat and pressure (up to several thousand Earth atmospheres), mixing the elements of the ocean and volcanic gases in a state-of-the-art high-pressure cooker and created a whooshing "pop" noise that showed the world the number of amino acids produced in such an environment.

Together they explored the implications of an inextricable life-and-Earth marriage, with no clear line of demarcation between them. The Earth was a restless, mutable globe in a constant state of change. Most of the critical microbial action was hidden from view. With that understanding, a new industry in bioremediation and bioenergy sprang up. Geology and microbial life made an intricate pairing that together offered clues to solutions of planet crisis. Suddenly microbes were getting more attention as researchers rushed to find ways to exploit their beneficial effects.

Microbes played a role in making soil, air, and the richest iron and gold deposits. They could help degrade industrial, nuclear, municipal, agricultural, and mining waste. What role had they played in past extinctions, and what could they do about those happening now and in the future?

What happened next changed our understanding of the potential of microbes in the world. It all began with a grade-school science report.

Three rivers

Growing up in Pennsylvania, the round-faced Andrew Knoll was heading to the library to prepare a report on the work of Lynn Margulis. There he came across her endosymbiosis idea. It fascinated the budding young geologist. "Everything I'd learned was Darwin's 'descent with modification,'" he recalled to me. "You're slightly different from your parents and they're slightly different from their parents. It really isn't part of the canon to have evolution occur because two organisms fuse into one. The idea that the reason an oak can photosynthesize is because one ancestor swallowed a cyanobacterium was stunning."

At the time the famed Harvard paleobotanist Elso Barghoorn was exploring remote cliffs and caverns featuring deep fossil records hidden in such distant areas as Cuba and South Africa. His insights made discoveries about the history of life tractable in a new way. Knoll followed those discoveries as a college student. What he did not know was that he would soon be accompanying Barghoorn on those trips.

"Ever since I was a kid, I have loved to hit a rock and see a fossil inside," he told me. "It was thrilling to hold something that was alive millions of years ago. Beyond finding a new fossil or making a new measurement, it is finding a new context, a record people didn't know about until fifty years ago!"

Knoll made it to Harvard and enrolled in Barghoorn's introduction to paleobotany. Barghoorn was pushing back the origin of life on Earth to 3.5 billion years ago, shortly after the end of the asteroid and comet bombardment, by painstakingly finding evidence of microbes in South Africa. Knoll would head back to the lab in Cambridge and analyze the specimens. "If you look at the earliest branch in the tree of life, 3.5 billion years ago," he said to me, "you could likely have a carbon and sulfur cycle that would have both bacterial and archaeal components." Of course, we do not really know if we can look that deeply with current evolutionary methods. "What made the bacteria interesting is that they have a number of groups or lineages that are photosynthetic," Knoll said. These photosynthetic cells evolved clusters of pigments of bewildering diversity that operated as tiny light-collecting antennas. There were at least six different groups of photosynthetic bacteria, and several of them did not use water as their main energy source. The very earliest photosynthetic systems were not even generating oxygen, and only the cyanobacteria did so later, somewhere between two and three billion years ago. At the beginning, the earliest life exploited any accessible source for their primitive metabolism—hydrogen gas, possibly reduced iron. Ancestors of those cells live today, some in mining and radioactive waste, some inside of us, proving that the most ancient metabolic pathways were anaerobic. Oxygenic photosynthesis, by contrast, was the "great ecological liberator," Knoll said to me. "It's what allowed you to spread life around the planet."

Moving into his own Harvard lab, Knoll led a team that analyzed rocks from different eras of Earth's history for the element molybdenum, which is sensitive to oxygen levels in the oceans. They tracked the oxygen levels on Earth through the eras and compared it to the known evolutionary history established through fossil records. High levels of oxygen encouraged evolution after the Great Oxygenation Event 2.4 billion years ago, which was why Knoll analyzed late Archaean basins from southern Africa and Australia, the mid-Proterozoic basins in Australia, and the Neoproterozoic successions in northern Russia and Australia. Microbial fossils in those places were detected by tracing the chemical compounds that create them. His analysis of 250-million-year-old rocks suggested that an early great extinction event, caused by excess carbon dioxide, paved the way for the dinosaurs, much as we humans were pouring carbon into the atmosphere from fossil fuels.

The analytical skills his team developed brought Knoll to the attention of NASA. In 2005 the Mars rover Opportunity was collecting samples and examining the planet's landscape, and Knoll was on the team working with the rover. The team discovered hematite formations on the planet's surface, confirming that water once flowed on the surface. The rover transmitted data for years after the battery was supposed to fail, and the mission was a big step toward understanding the history of the planet that a growing number of scientists thought made a great bet for Earth life's origin. But back on Earth, a science "pop" was breaking open a new vision.

The hiss and "pop"

At an innocuous, neatly organized white-walled lab of the Carnegie Institute's Geophysical Department, Robert Hazen gingerly pulled a golden tube from a metal cylinder, where it had been cooking at the pressure of two thousand Earth atmospheres. The enormous, gas-loaded contraption looked like something out of a science fiction movie. His graduate students ducked.

In a leafy neighborhood of Washington, D.C., Hazen's team was recreating the steps by which life might have started at deep-sea volcanoes. In these oases on the shifting ocean floor, teeming communities of giant tubeworms, lobsters, and sulfur-loving microbes had lived for eons without any human knowing of them. Then Günter Wächtershäuser suggested life had started down there. In his lab, Hazen recreated the vents' high heat, anaerobic conditions, and unbelievable pressure, mixing the elements of ancient ocean water and volcanic gases. As the tube opened it created a whooshing "pop" and oozed a yellow-brown oily goo showing the number of organic compounds produced in such an environment. "You expect nature to be orderly," he told me. "But no, nature is sometimes very messy!" They made many more organic compounds than they had expected. "The problem," he recalled, "was winnowing them down."

Robert Hazen was born in 1948 and was an active trumpet player since childhood, leading to a lifelong interest in marching bands and musical appearances with the National Symphony, the Metropolitan Opera, and the Washington Chamber Symphony, as well as in numerous recordings. It also compelled him to communicate his science. "Musicians tend to be very intellectually active and curious," he told me. "They'd ask me about my work. I'd have to describe my work to laypeople and try to make it amusing and entertaining. That was how I began writing about science."

Hazen received his bachelor's and master's degrees in earth science at the Massachusetts Institute of Technology in 1971 and his doctorate from Harvard University in the field of mineralogy and crystallography in 1975. His postdoctoral fellowship at Cambridge University led to an opportunity to work on high-temperature superconductors. He kept a diary of the high-stakes science and turned it into a first popular book, *The Breakthrough*. Together with the George Mason University physicist James Trefil, Hazen later developed a course called Great Ideas in Science, "which really got me to step back from my specialty . . . and think about what are the big unanswered questions. I wasn't really addressing any in my research and life is short, you want to have an impact."

Since 1996 Hazen had felt that any plausible origins scenario must be consistent with geochemical constraints. He developed a new approach called mineral evolution, which investigated the coevolution of the geosphere and biosphere. Hazen and his team used "pressure bombs," metal containers that pressurize and heat minerals to the equivalent of the environment at the ocean bottom, to test if this type of setting might be where life began. These "pressure bombs" were extremely dangerous—if something were to go wrong, the explosion would destroy a large part of the building. Therefore, the user operated the experiment behind an armored barrier.

During Hazen's first experiment, his small concoction of water, pyruvate, and carbon dioxide produced a yellow-brown goo containing thousands of different organic compounds. Later experiments proved that the combination of nitrogen, ammonia, and other early Earth molecules yielded many organic molecules, including amino acids and sugars. These experiments showed that the basic molecules of life certainly could form around volcanoes under the ocean at hydrothermal vents.

Hazen gave highly prized public lectures on his findings. He stood straight, enunciating with a command that stemmed from his years as a professional trumpeter, and encouraged his colleagues, such as University of Tennessee's Linda Kah, to talk with reporters. Outside of a Vancouver Geological Society conference room he met Kah, who was with the Mars Curiosity team, and introduced me. Hazen studied the Earth minerals shaped or created by microbes. He also collected trilobites and brass band artwork (some of his collections are on display at the National Museum of American History and the National Museum of Natural History) and edited an anthology of eighteenth- and nineteenth-century poems about geology. He was moving on to a larger study of complex organization of systems in science, much as Woese ended up studying the complex organization of cells. "The experiments we're running now," he told me from his home near the Chesapeake Bay, "show quite definitively and remarkably that you get huge enhancements of stability in amino acids when you have mineral surfaces and a very reducing environment, that is, one that has hydrogen in solution."

But Hazen wanted more—to understand how the patterned, emergent complexity of microbes and the diversity, distribution, and disparity of minerals related to each other and a whole array of other parameters, such as climate and volcanism Past researchers could make a grand narrative of life's emergence, but he wanted to do so quantitatively, and some of the mineralogical equations he came up with, for the statistical distribution of minerals over the Earth's surface were "exactly the same as ones you would use in biology," he said. He analyzed the many ways microbes created and shaped Earth's minerals, from tissues like bone to carbonates in coral to copper, iron, and gold deposits. He sent me a draft of a paper on "chance and necessity," echoing the famous phrase of the Nobel Prize winner Jacques Monod in his book on evolution. Hazen was using the large number of random events (LNRE) theory, in which he and his team determined that in the entire universe, only Earth has precisely the minerals it has. When he presented it at Vancouver, a tweeter noted, "I love when Robert Hazen's talks." But the idea had not yet caught fire.

Energy drinker

The Rhine River runs through the industrial heart of Switzerland and Germany in a coal-rich valley once filled with toxic smokestacks and factories. A possible reason Germany, France, England, and the United States all experienced the industrial revolution is that the four countries contained coal deposits to power the first, crudely inefficient steam machines. Early life might have done much the same thing with natural energy sources in hot pools or explosive volcanic vents. That insight set off a rush to discover energy gradients in mineral springs from Kamchatka to Yellowstone, igniting a revolution in our understanding of the possibilities of microbes to shape our energy needs.

From black holes to colliding stars, the universe features special sites where energy is much higher on one side of the equation than the other. An exploding supernova, a comet hurtling toward its sun, a volcano undersea or above the ground, even the heating and cooling of a tidal pool could create the inorganic imbalance on which life would seek to capitalize. Forget what you thought you knew about primordial Earth. The early Earth was, in Hazen's view, in fact wet and relatively pleasant. The atmosphere stank with ammonia, sulfur, and weekly meteor and lightning strikes but would have generated tons of amino acids. Of all the definitions of life, the University of London's Nick Lane argued in wittily written popular books and articles, the one that mattered most was energy.

At first, many thinkers thought the remarkably large protein complexes that catalyze life's cascade of energy reactions could not have survived in the oceans, Lane said. Then he and others had a revelation.

Late in 2012 Lane popularized the idea that hydrothermal vents make batteries composed of protons. What these precise structures revealed were clues in the mineral centers in life's vital enzymes: the iron sulfide common at vents is also found at the heart of the respiratory enzymes that catalyze electron transfers. Lane and Heinrich Heine University's Bill Martin argued that life began in the alkaline hydrothermal vents, like those near the Mid-Atlantic Ridge; these were more temperate than active volcanic vents, structures discovered only in 2003. "This is really cool, novel stuff," the University of California's Jan Amend told LiveScience of the new theory. Not a miraculous bolt of lightning but a simple, everyday source of power out of vents that lasted for hundreds or thousands of years had provided the battery that started life.

The hydrothermal-vents hypothesis was still opposed rigorously by Stanley Miller's original assistants, Jeffrey Bada and Antonio Lazcano, who argued that RNA in water falls apart too easily to be formed at the ocean bottom. Bada organized the Miller papers, as well as the original 1952 Nobel Prize–winning contraption, in a closet at his lab at the University of California, San Diego. He used the contraption to brew coffee.

Hazen responded with a gusto born of being a professional musician. "That may be true in a test tube," he said to me at a Vancouver conference of the Geological Society of North America, "but was it true in a volcanic vent? Because you have all kinds of surfaces, you have nanoparticles, you have chemical complexity absent from a test tube." He showed you could get huge enhancements of stability of amino acids when you have mineral surfaces and a very reducing environment, all under high pressure. He pursued the idea that minerals and life form a dynamic, unfolding stream of interrelationships. Early life may have formed in tandem with gold and pyrite, and several minerals in turn formed in tandem with life. Another beautiful blue mineral, called ringwoodite, revealed that a hidden aquarium exists three hundred miles beneath the Earth's surface, and that it may hold as much water as all the Earth's oceans together. "We're seeing organizational patterns that reflect a continuity among systems," Hazen said, "that a decade ago were considered quite different."

But what did that mean? It meant that the life's emergence might be considered a phase transition that manifested itself as a change in how chemistry processed and used information and free energy. Understanding that phase transition required a new approach to non-equilibrium physics—from rocks to life.

As the former students of Knoll and Hazen scoured the remote regions of the Earth for the world's first evidence of microbial life a host of industries, old and new, new microbial industries sprang up—heavy metals, petroleum and mining waste remediation, plastics and food-waste processing, polyaromatic hydrocarbons (PAHs), and even greenhouse cleanup. Microbes were the major players in carbon and nitrogen fixation and methane, iron, and sulfur metabolism. They offered the potential to clean up our mess.

India took a leading role in the commercialization of microbes for bioremediation. With pollution a world crisis—a ten-square mile bag of plastic floating in the Pacific, carbon emissions, and petroleum spills being the most important organic pollutants and produced in amounts ranging from one to eight million tons a year—a dozen companies and researchers such as Surajit Das, Pasyap Kumer Dubey, Punit Kumar, and others, sought to uncover the ways in which microbes recovered soil destroyed by mining and radioactive waste, petroleum and heavy metals runoff, and fertilizer and asphalt overuse and misuse. Hundreds of companies went into the wild looking for the many unknown capabilities nature had nurtured.

Microbes produced chemical reactions that transfer electrons away from contaminants, rendering them benign or even useful. Take plastic waste products: Mangrove swamp–associated bacteria like *Micrococcus*, *Streptococcus*, and even *Staphylococcus* can degrade plastics. For cleaning petroleum spills, *Acinetobacter, Marino-coccus, Methylobacteria*, and others were useful. Oil seeps in the Gulf of Mexico and the Persian Gulf created the thriving communities of bacteria, Archaea, and sea worms, some of which grazed on methane, that Samantha Joye had observed. Thousands of the natural seeps were significant enough to be visible from satellites, leaking 2,500 barrels of oil per day in the Gulf of Mexico alone.

Another thriving industry was in the genetic improvement of common microbes such as *E. coli* to serve as bioremediators of toxic spills of PCBs, benzene, and toluene, a by-product of industrial production. Such techniques included bioventing (injecting oxygen around a spill to spur bacterial growth) or bioaugmentation and biopiling (injecting microbes directly or removing spills to a different site for bacterial decomposition). The old,-well-known practice of composting made for a growing trend in backyards across America and Europe. As of 2012, in the United States, about 35 percent of municipal solid waste was recycled or composted, with the balance sent for either energy recovery (11 percent) or to municipal landfills (54 percent).

A microbial energy industry sought to use the waste of landfills and agriculture to generate electricity. A microbial biorefinery industry was seen for cyanobacteria and microalgae that could be grown in wastewater to create food, fuel, and energy. Energy was the focus of a new interest in microbial electrochemical systems. Some microbes could be used to reduce carbon dioxide levels. Bioreactors or fuel cells powered by microbes spawned a series of start-ups in India. Microbes could also clean up the waste of fish farms and clean our drinking water, as they did in the wild.

The more straightforward and successful industry was microbe remediation of agricultural, oil, and heavy metal waste, as well as repair of mold damage. The Oakland-based company Anaerobe took remediation further. Anaerobe sold microbe blends that could rapidly turn tons of agricultural and invasive plant waste into fertilizer and hydrogen fuel. Anaerobe's founder, Mike Cox, was a Texas-raised ex-Marine and Stanford hospital lab technician and inventor, whose daughter, Laurie Cox, was at Harvard studying the gut microbiomes of the obese or elderly and sufferers from multiple sclerosis. When Mike Cox realized that an astonishing 40 percent of resistant hospital biofilms could not be analyzed reliably, he designed a rapid, highly accurate "Cox detector" that funded his microbe remediation business. He received grants from the Department of Agriculture and from Silicon Valley but struggled to expand the business. "It is a gold mine," he explained to me. "The problem is getting governments to believe in it."

As the ideas of microbial bioremediation caught on around the world, two former Knoll students and Hazen's colleague Nora Noffke took off to find microbe sediments in Africa and Australia. The Knoll student and NASA researcher Linda Kah did the same in Africa. Suddenly the microbes were getting more attention as researchers sought to find ways to exploit their beneficial effects.

The race was on to harness and utilize those microbial technologies. "If there's a paradigm shift, there's a recognition that the geosphere has a profound negative and positive feedback with the biosphere," said Hazen. The physicist and biotech entrepreneur Robert Carlson predicted a new era of biological engineering. A few early investors and companies became interested, and annual student conferences solicited ideas for bioengineering new microbes to confront our energy and remediation needs.

The most explosive series of discoveries suggested that microbes alter our moods and emotional health. One place people looked for that was in the dirt.

In the Garden

Microbes, Power, and Health

It is Mother's Day in our Chicago suburb, and I am helping to plant the flower planters — or, rather, staying mostly out of my wife's routine. I wheel the rusted red wagon around the greenhouse, piling on trays of new flowers, thinking about each one liberating energy from sunlight and water, burning the hydrogen in water for power, and releasing the oxygen as waste. The car is packed with soils, plastic trays, and the rich loamy scents of bright-colored gardenias, geraniums, begonias, ficus leaves, and spikes. A soft spring breeze carries the moistness of spring clouds and the pulpy smells of earth.

Every spring millions of weekend gardeners work the backyard soil, pulling out window boxes and laying out geraniums, vincas, daisies, zinnias, begonias, potting soil, herbs, and vegetables. On city balconies and vacant lots, in stretches of farmland or high in the mountains, many of us find a kindred spirit with the Earth. The sense of well-being feels a bit miraculous but is entirely natural: The gardener's feeling of "wondrous life," to quote the poet Andrew Marvell, has hard science to back it. A handful of soil contains billions of microbes, more than any other handful of any other region on Earth. Some of them are natural mood enhancers and more. Many antibiotics have come from soil—and these critical medical healers may have to help combat the crisis of antibiotic resistance.

Most people view soil as something to be washed, beneath, old, or "boring as dirt." But soil is vividly young, alive, and dynamic, adapting to and improving the environment. Soil microbes help feed the plants that recycle our air, purify our groundwater, and decompose dead animals and plants, all without a trace of human interaction. Most beguiling is to see soil in the images of Mars, and many other places in the solar system, wherever there is a rock—on asteroids, planets, and the moon, where it's called regolith, or comets. On the comet Churyumov-Gerasimenko, the Rosetta probe showed rubble looking like Rocky Mountain scree, but older than the solar system itself.

Soil is home to a menagerie of worms, insects, fungi, lichens, and, most of all, microbes. A single gram of soil houses seven billion microbes, more than all the humans on Earth, and a typical backyard has a trillion billion useful microbes. Soil houses more organisms than there are stars in the universe. Vancomycin originated from a bacterium found in soil, while penicillin came from a fungus that molded bread and rotted plants and fruits. The antibiotic streptomycin and twenty-one others were discovered in one soil microbe. The immune suppressor that makes transplants possible was discovered in the soil of Easter Island. A third soil compound was shown to stimulate the natural antidepressant serotonin, improving cognitive function and reducing inflammation. Perhaps that is why some people find gardening so relaxing.

The discovery that microbes may affect our physical and mental health sparked a gold rush based on a new understanding, harking back to an old idea called the "hygiene hypothesis," that dirt helps make us healthy and happy. Exposure to microbes may help protect us from depression, allergies, asthma, and autoimmune disorders. It was no mere academic quest. Resistance to antibiotics was a worldwide crisis. Many emotional and mental illnesses were on the rise — autism and depression, obesity and immune disorders — and many defied conventional treatments. Beguiling evidence, pioneered by small labs outside the mainstream of research, suggested that clues to the solution lay in microbes. Yet the explosion of knowledge had only identified the genes of some 0.01 percent of the Earth's smallest creatures.

As I laid out the heavy, peeling flower boxes in the yard, a dozen research labs were pursuing that goal. The question was who, and how.

In the wormery

The Illinois ravines run for miles as I walk along with my retired brother-in-law, a gentleman farmer, near his compost heap. Wind whips down, and the fallen leaves rustle. Walk in a backyard, farm, city lot, or remote mountain hillside. The dirt you see is the result of profound geochemical and biological processes. On Earth, soil is constantly changing through the processes that create it. It is made through roughly five such processes: rain and wind weathering, plant decay, and the metabolism of microbes of the guts of earthworms and grazers such as cattle. The process is slow and precious: it takes fifteen years to make half an inch of soil in a wet climate, even longer in a dry climate. The soil contains the nutrients and water plants depend on. The Dust Bowl of the 1930s and the current fragility of drought-plagued African, Mongolian, and Australian agricultural land show how losing it leads to ecological disaster. Soil and the microbes that make and inhabit it are the basis of life.

Because of their abundance, microbes in dirt are in a constant battle for nutrients. That battle can spark various devious and complicated chemical signals to foil each other, meaning that soil microbes have a far wider array of chemical abilities than we do, which is what made them a main source of antibiotic medicines back in the 1950s and '60s. Some researchers argued they could produce yet more inexpensive medicines, energy sources, and radioactive and sewage waste remediators. But we knew very little about them. Tens of thousands of unknown soil microbe species lay beneath our feet, with huge variations from region to region and with such potential "the microbial community in the ground is as important to our health as the one in our guts." They help provide plants with nutrients and resistance to pathogens and predators, help plants to signal each other, sequester the Earth's carbon, filter our water, and protect us against drought and flood. They offer such valuable economic services that a new soil microbe industry was valued at $1 billion in 2012 and growing rapidly.

The earthworm makes our dirt. Charles Darwin was the first to assess the impact of the guts of earthworms on the creation of the English garden. While he did not know the incredible diversity of microbes in soil, he correctly saw that worms were critical to making it, so much so that he spent years calculating the rate of soil creation in England. He constructed a "wormery," because that is a thing, where he observed earthworms eating plant and animal matter and excreting a nutrient-rich base for soil. He calculated the rate and production of England's total soil mass made by its worms. "Long before [humans] existed the land was in fact regularly ploughed, and still continues to be thus ploughed by earthworms," he wrote. "It may be doubted whether there are many other animals which have played so important a part in the history of the world."

But what about on a smaller scale? An unfathomable number of microscopic organisms live in the dirt they helped to create. Each has its own role, most of which we do not yet understand but without which most plant life would not be possible. These include the familiar cyanobacteria that, along with the plants that incorporated them as chloroplasts, create the Earth's oxygen. Add to that the natural decomposers of dead tissue that save us from a planet of animal corpses and the extremophiles of hot vents and oceans, cousins of the same microbes in our gut that help digest our food and attracted such attention, and even some of the fungi that ferment our wine and beer.

Soil microbes fed plants, cleaned our groundwater, provided a foundation for our homes and our games, and made many of our medicines. Soil is a perfect laboratory for understanding survival. One group of soil microbes alone, actinomycetes, was the source of some twenty-two different antibiotic compounds, including actinomycin, neomycin, and the widely used streptomycin. An Easter Island soil microbe gave us a critical immune suppressant. Dozens followed. But once resistance to our common antibiotics had reached the level of a global crisis, new researchers raced to understand the healthful role of such microbes. Then came a soil microbe that changed our view of emotional health.

Norwegian wood

Susan Jenks was a medieval-studies major at Vassar College in upstate New York who moved to another field, psychology and emotion, and from there to the brand-new field of the genetics of behavior. She was hooked and switched to a science major. But for years she studied animals, engaging in fascinating research in the northwest on the social behavior of timber wolves with Ben Ginzberg until she came across the English laboratory pioneer Chris Lowry.

Lowry was at Bristol University studying microbes found in mud, untreated water, and fermenting vegetable matter, *Lactobacillus* and saprophytes. *Lactobacillus* was a star of the probiotics movement, helpful in treating intestinal issues like preventing diarrhea in travelers or chemotherapy patients, and those suffering from irritable bowel syndrome. Saprophytes were the microbes and fungi that lived on decaying or dead things. Lowry, however, focused on another obscure microbe, called *Mycobacterium vaccae*, a natural soil microbe first isolated in cattle dung in Austria, for which it was named. This group of *vaccae* were interesting as the microbes related to inflammation. Europeans and Americans had higher levels of inflammation than people from rural, undeveloped environments. Because depression and other psychiatric disorders, it was seen, were associated with a mild elevation in inflammation, Lowry wanted to know if *M. vaccae* affected emotional health. He attached heat-killed *M. vaccae* to a vaccine he injected in mice. The mice showed marked improvement in their willingness to attack a maze. "This seemed to me to be a fascinating area no one was investigating," Jenks told me from the garden of her home in upstate New York.

Jenks was an avid gardener. She and her colleague Dorothy Matthews were teaching at an undergraduate institution, SAGE Colleges, and casting about for an overlooked area combining their interests in psychology, behavior, and the natural world, where they could compete with more well-heeled researchers. "We didn't have major grants," she recalled. "What we did have was access to mice and to a new technique."

The technique was a cheap form of microbial gene sequencing, coupled with their long experience in psychology and neurobiology experiments in mice. They began by feeding healthy *M. vaccae* to one-third of their mice, then making the mice run a maze. In the cramped, fluorescent-lit facility in their small upstate town, those mice with the microbes did much better than mice without the *M. vaccae* in their feed. Jenks and Matthews sampled the mice's blood and found that *vaccae* caused cytokine levels to rise, which increased the production of serotonin. They thought the microbe affected the structure and expression of genes, but not their makeup. However, they were not gut microbes. Jenks and Matthews hypothesized that soil *M. vaccae* could be inhaled and work through the lungs into the bloodstream. It was not a gut microbe, but rather a commensal organism, having an external effect on the body. "Everyone in the media seemed to misunderstand that," Jenks told me.

When they published their findings in a small journal and presented them at the 2010 annual meeting of the Animal Behavior Society, the story ignited. "Can Gardening Help Your Emotions?"

queried the *Atlantic*. "How Dirt Makes You Happy" was the headline on the website Gardening Know How. An earlier headline in *Discover* asked, "Is Dirt the New Prozac?" "How the Bacteria in Dirt Make You Happier and Smarter," said the Healing Landscapes website. Those microbes came to be one of the biggest arenas of the rise of the twenty-first-century illnesses affecting mental health. In 2013, Jenks followed with another study showing mice fed *vaccae* also experienced improved learning and lessened anxiety. To belong to the microbial club took special abilities—to live with or without oxygen, attach to the "right" habitat, evade bacteriophages or viruses that attack microbe, and appease a reactionary immune system, to change fast, grow rapidly, and resist stress when jumping to another host.

The "smart part" was a finding of researchers at University College in Cork, Ireland, who had pioneered the field. These included Ted Dinan and John Cryan, who determined that mice fed *vaccae* from soil had improved cognitive function. That same year, Swedish scientists saw that the gut microbiome regulated brain development and prevented depression. As dozens of researchers raced to understand how *M. vaccae* worked on mice, one who had suggested as far back as 1990 that microbes played a key role in our mental health became famous—New Jersey–born Mark Lyte.

"What the hell has happened?"

When the excited Lyte gave his first presentation at the meeting of the American Society of Gastroenterology in New Orleans in 1990, exactly two people sat in the audience. One was his lab assistant. The next time he spoke in the same city, at a Field of Infectious Diseases gastroenterologists' meeting, 150 showed up—but no one asked a question. As he was about to beat a fast retreat, the group of doctors mobbed him. "Hundreds of our patients on antibiotics complain about becoming more anxious," one said. "You mean they're right?"

Mark Lyte was born in 1955 and grew up in the small working-class suburb of Bergenfield, New Jersey. He began as a medical technician taking blood in a city hospital, fell in love, and took his girl-friend to the Weizmann Institute in Israel to escape his parents' disapproval. They kept asking him to become "a real doctor," but Lyte was on his own path—first clinically focused, then highly research oriented. Beginning in Minnesota in the 1990s, Lyte studied the immune system's response to stress. He conducted experiments showing that stress hormones such as norepinephrine sparked blooms of bacteria in humans. Then, in 1998, he showed that mice injected with pathological bacteria were more cautious and risk-adverse than control groups. Stressful condition for calves worsened infections.

Few scientists noticed. But an outbreak of *E. coli* in the drinking water of the small Ontario town of Walkerton made for a real-life experiment. After one year of contamination, the rates of irritable bowel syndrome rose precipitously. So did something else: the number of Waterton residents diagnosed with clinical depression shot up by 350 percent.

In Cork, the teams of Ted Dinan and John Cryan were studying the apparent connection. They injected germ-free mice with the microbiota of obese mice, and the animals gained weight, some of them showing less interest in completing a maze. When Washington University's Jeff Gordon and Wisconsin's Margaret McFall-Ngai started to back up Lyte's general observations with hard data, the National Institutes of Health and the National Science Foundation made Lyte a finalist for a Pioneer Award. But a reviewer asked: if killing microbes causes anxiety, why weren't we all depressed?

The fact was, seemingly, we were. The rate of incidence of autism, across the broad spectrum of its many levels of effect, was skyrocketing, as were rates of depression, anxiety, and obsessive-compulsive disorder. At Washington University in St. Louis; at Texas Tech, where Lyte worked; at University College in Ireland; at McMaster in Toronto; and at Caltech in Pasadena, researchers like Paul Peterson and Sarkis Mazmanian sought to understand why. We each have more than a thousand different types of bacteria that live in our digestive tracts, helping us to break down food and absorb nutrients. But when we take antibiotics, the drugs kill many of the healthy intestinal flora that help us digest food. That was Lyte's message to the gastroenterologists in their New Orleans meeting.

At the same time, neuroscientists at Cambridge and Oxford Universities found that *prebiotics*, the nondigestible fiber that acts as food for good bacteria, also had an antianxiety effect. Thirty percent of patients who take antibiotics report suffering some form of gastrointestinal distress, so doctors commonly prescribe taking probiotics to repopulate the digestive tract with healthful bacteria.

Ignited by books, magazine articles, news segments, and podcasts, interest skyrocketed. The idea that microbes are allies that protect us and other animals from harm was popularized by the well-regarded new science writer Ed Yong in blog posts for the *Atlantic* and a *National Geographic* series titled "It's Not Rocket Science." Yong compared human microbes to animals in a national park, which help train the park rangers—the immune system's defenders—who patrolled them. When Lyte was profiled in the *New York Times Magazine*, it became the most downloaded article of the publication. Iowa State University offered Lyte more money and a complete, dedicated lab floor to move there from Texas.

One day before Lyte left Texas, he was working out on the stationary cycle with his wife at the university gym. On TV he saw a *Dr. Oz* episode pitching gut microbes to prevent Alzheimer's. "What the hell has happened?" exclaimed Lyte to his wife, who told him to shut up as the Oz television episode drew others to the screen. "We don't know what the good bacteria are!" he whispered. "We're not saying we can prevent Alzheimer's!"

Around the world, researchers set to work. What was needed was better evidence to separate cause from effect.

A Gang of 20

The early 2000s marked the triumph of the human genome sequence. Tools such as automated gene sequencers, mass spectrometers, and data analyzers helped interpret the beautiful colored wheel diagrams that depict human gene sequences. But although some 70,000 human genes had been projected, only 25,000–30,000 were found—about the same number as the worm *C. elegans* had. Estimates of gene numbers were fuzzy because the definition of "gene" was fuzzy. Where were some of the missing 40,000–45,000, and what was doing their work in the body? some researchers asked. Perhaps some of that expected work was being done by genes from the human microbiome. A consortium of institutions turned to the microbiome to answer the question: Was there a core of human gut microbes that guarantee our health, as opposed to the handful that cause infectious disease? The U.S. National Institutes of Health multi-institutional Human Microbiome Project released its early findings in the summer of 2012 and continued to refine them.

Elaine Hsiao was a "typical Los Angeles high school student," she recalled. Her father was a poet and musician who died early, and she gravitated to the sciences. As a graduate student in biology, she entered the lab of Caltech's seminal thinker Paul Patterson in 2007. Patterson was studying mouse models for autism, schizophrenia, and such neurodegenerative diseases as Huntington's chorea and was turning to the animal's microbiome for clues to its emotional and mental health. The microbiome and immune link was becoming established through Jeff Gordon's work at Washington University, showing how gut microbes affected the weight and health of both Africans and obese Americans. At the same time, in Ireland John Cryan and Ted Dinan had formed a major group studying the role of microbes in mouse mental health. In Canada Evan Collin at Toronto's McMaster University was doing much the same. American researchers, such as Caltech's Sarkis Mazmanian and Iowa State's Mark Lyte, were linking microbes in our bodies with the emotions and feelings in our hearts and minds by analyzing how our microbes were protecting or augmenting our metabolism, immune protection, growth, and emotional and physical health.

This small, disparate group of pioneers created the techniques and technologies that showed the human gut microbiome featuring a cast of a thousand but dominated by two phyla, Firmicutes and Bacteroidetes. These microbes aided in our nutrition, immune system development, and fitness. It was suggested that some large percentage of children diagnosed with autism or depression and anxiety had some sort of feeding-related concern.

They were following the neuroimmune contribution to autism, "a new literature on bacteria that promote immunosuppression," Hsiao told me. "If you took autism mouse models and treated them by bone-marrow transplant, their behavior improved." The next question was, instead of bone-marrow transplants, could they use microbes? The behavior and emotional health of those mice improved. Now the question was, could we look into ourselves for new therapies? In the spring of 2015 a study out of the Leiden Institute of Brain and Cognition at Leiden University in the Netherlands suggested that probiotics aided in improving mood. The researchers examined forty healthy young adults. Every night for four weeks, half consumed a powdered probiotic supplement, Ecologic Barrier, containing eight types of bacteria, including *Bifidobacterium*, *Lactobacillus*, and *Lactococcus* (these types have been shown to mitigate anxiety). The other half took a placebo. Those on probiotics felt more improvements in their moods than those on placebos.

By 2016 Hsiao had identified twenty spore-forming bacteria that helped increase serotonin levels in the mouse gut. The new generation of researchers, including Hsiao and Harvard's Emily Balskus, employed next-generation gene and protein analysis as they sought to identify specific types of bacteria, and the compounds they created, that modulated human serotonin levels to figure out the biochemical pathways to use and to better understand the links to human disease. Reviewing Hsiao's work, the National Institute of Allergy and Infectious Disease researcher Yasmine Belkaid speculated that the subtle interaction of gut microbe–induced short-chain fatty acids and serotonin levels constitute the human body's "second brain," which in turn helped shape the immune system and affected emotional well-being.

As some of these new scientists formed companies and established pharmaceutical companies scoured the ocean floor, the work raised questions about what microbiome science is and does in our time. Were the effects caused by changes in microbe populations, or were the conditions changing microbes? Was the growing public interest driving a bias toward positive results? The answer was, probably a little bit of both. In some cases, however, some of the claims based on the facts spun out of control.

Then came a setback.

Uncharted territory

It turned out that the number of microbes in the human body was far fewer than the hundred trillion advertised. It had always been a wild guess, or rather a lucky one. It all began in 1970 when microbiologist Thomas Luckey did a quick calculation, and decided we housed some one hundred trillion, ten times the number of human cells.

But no one had ever really tested the hypothesis. By 2014 an Israeli team had corrected the number—to between twenty and thirty trillion, about the same as the number of human cells, and varying widely in population during the day. In the morning, they said, we house more microbes—because as we sleep, the anaerobes grow in our mouth. After a toothbrushing and a couple of trips to the bathroom we house far fewer, and we might lose a third of our gut travelers in a bowel movement. The other issue is that microbial cells were far smaller than human cells. We are, indubitably, human.

Still, the number of microbial cells was huge, and the number of microbial genes in the human body was much larger than the 25,000 genes we housed. What was needed was better hard data. Back in 2012 an international team of microbiologists came up with a first-ever catalog of microbes in the body: The sample came from eighteen places on the body. Scientists identified some 10,000 species of microbes in the body, including many never seen before. "This is like going into uncharted territory," the Jackson Laboratory's George Weinstock told National Public Radio. Those 10,000 species possessed some eight million genes, roughly twenty times the figure used as the number of human genes.

Whatever the number, microbes were becoming one of the biggest topics in life and environmental science. One could not open a website, article, or Twitter update without seeing a new claim. The problem was finding ways to prove the claims.

The microbiome redefined the ways we consider the human self and our bodies, and a burst of companies sought to capitalize on the revolution. The Caltech medical microbiologist Sarkis Mazmanian started two companies, one with Hsiao they cofounded for autism work called Neurobiotics, the other for brain disorders and Parkinson's, among other maladies, called Axial Biotherapeutics. Mark Lyte had two companies. John Cryan and Ted Dinan in Ireland had a new company. Northeastern University's Slava Epstein started a company called Novobiotic. The hygiene hypothesis, encouraging young children, especially girls, to play more in the dirt, began to take hold.

With the advent of capital came some hype about probiotics and misinformation about the science backing them. "Digestive bliss," offered one probiotic's website. "Thirty Billion Colony Forming Units (CFU) of 'good bacteria' in every probiotics capsule," offered another. *Lactobacillus* and *Bifidobacteria* were the most popular microbes in these dietary supplements, and they seemed most applicable to treating irritable bowel syndrome. But the FDA had not yet approved any for any medical treatment, and the supplements often did not contain exactly the constituents advertised.

At the same time, an alarming figure for the number of microbes on the New York subway, at some billion per inch, had to be corrected. The figure for the number of microbes in the human body, outnumbering human cells by ten to one, turned out to be based on a very rough estimate by an obscure researcher. The proper ratio, more like one or one-and-a-quarter to one, fluctuated a good deal depending on time of day.

The University of California, Davis evolutionary biologist Jonathan Eisen called the hype "microbiomania" and became a widely read watchdog on the topic. What was needed was more hard evidence as researchers sought to translate science into application. New microorganisms, for instance, could indeed help address the problem of antibiotic resistance, but the going was tough and arduous. For example, the treatment of choice for resistant infections, vancomycin, originated from a gram-positive bacterium found in soil and, with the ongoing battle of bacteria developing a heightened resistance to our current antibiotics, scientists returned to searching for similar strains in new soils.

During one of these searches, a soil sample taken in Maine was found to contain a bacterium capable of targeting methicillin-resistant *Staphylococcus aureus* (MRSA) and *Mycobacterium tuberculosis*. Their lead researcher, Slava Epstein, had devised an ingenious tool that could isolate, identify, and grow potentially useful new strains much more easily than in the past, laying the groundwork for a "revolution in microbial ecology," he told me — "microbial biology without the microbiologist." His team in Greenland found vastly different microbial ecologies in various patches of similar soils. He called the discoveries a whole new ball game.

Another sample was found to contain *Mycobacterium vaccae*, the bacterium that interacts with neurons in much the same way as the antidepressant Prozac does, with effects comparable to those of antidepressants via the release of serotonin.

Research turned to the secondary metabolites of microbes, the compounds they produce, as in the lab of Harvard's Emily Balskus, as microbes showed the potential to address global challenges. Balskus was interested in the structure of the natural molecules made by living organisms for drugs. Their "beautiful architectures were very inspirational to me," she recalled. "The major challenge is going from a descriptive understanding to a mechanistic understanding to improve our ability to look at sequences and infer real information about function, about what the microbes are doing and not just who they are." She and others explored the logic of what they called secondary metabolism, the by-products of digestion, to understand the working of secondary metabolite gene clusters. In Norwich, England, the molecular biologist David Hopwood cloned the first biosynthetic gene cluster and raced to figure out how streptomycin is made by soil microbes and their secondary metabolites.

Three places where yeast microbes had been understood since before the times of Homer were in bread, wine, and beer. Research entered a whole new phase as brewers and bakers came to understand and drive the precise logic of microbes' secondary metabolism.

This one's for you

On a cool summer night in Chicago, far from the science bonanza, I wander into the pub called Begyle Brewing in a north-side factory brick building. The twenty-four-year-old chief brewmaster Liz French, who has a Ph.D. in microbiology, meets me at the back, wearing a black T-shirt, jeans, and work boots. Begyle is a community-supported business founded in 2012 as one of the new movement of neighborhood fermenters making up some 15 percent of the American beer market, with forty-eight in Chicago and up to a hundred in the region. It is a spare, dog-friendly place on a wooded side street, your friendly neighborhood bar if the bar was in a former factory and artist's gallery, with a zinc counter and Skeeball lounge, and a clientele of tattooed women and bearded men. "It's big on Father's Day," French says.

To a home brewer, there is nothing like watching a fifty-gallon clear jug of yeast, hops, barley, and water doing its microbial bubble and swirl for two weeks or so, generating so much carbon dioxide that the gas must be siphoned through a plastic tube into a holding tank that itself bubbles like a cauldron. Virtually every American city features a host of microbreweries like Begyle, each with its own signature—some playing vinyl records, some with vintage guitars as decoration, often in nondescript abandoned small-goods factories whose jobs long ago went overseas or to robots.

After French received her doctorate at Ohio's Miami University, where she studied lake microbes, her microbiologist boyfriend was awarded a postdoc at Northwestern University. She moved with him to find her own postdoc. As she interviewed, she would stop into the struggling Begyle, founded by her boyfriend's roommate. "The brewmaster kept asking me questions," she recalls. "He was in over his head." She explained to him the trickiness of keeping wild yeasts out to prevent them from spoiling the fermentation, in which the small fungus consumed oxygen until anaerobic conditions were achieved. During anaerobic growth, the buds transform maltose and other fermentable sugars, giving the beer its taste and kick. She answered so many questions that finally the owner asked if she wanted the job.

To maintain the quality of the yeast for up to sixteen generations, she peers through her microscope day after day, without the benefit of a giant company's gas chromatograph for flavor, a spectrophotometer for detecting acid compounds, a liquid chromatograph for measuring hop resins and sugars, or a near-infrared analyzer. "I never thought I'd use my Ph.D. for this," she said.

Other industries based on microbial functions also took off in the growing international fermenter subculture—home cheesemakers, yogurt cooks, and sauerkraut aficionados, linking centuries-old recipes with twenty-first century science. We have known since the early 1900s that the roots of plants were sites of great microbial activity. Bacteria found in the soil performed a process called nitrogen fixation, in which the inert form of atmospheric nitrogen is converted into a more reactive version that can be used to build, sustain, and propagate plants, and thus almost all visible life that depends on them. A handful of new companies sought to reproduce nitrogen fixation as an alternative to fertilizer.

Microbes could produce inexpensive chemicals, and several companies raced to exploit that trait. The George Church lab at Harvard's

Wyss Institute of Biological Engineering devised a way to make *E. coli* thirtyfold more efficient than previous methods in producing chemicals. "We make the bacteria addicted to the chemicals we want them to produce," Jameson Rogers, of the Graduate School of Arts and Sciences at the Harvard School of Engineering and Applied Science, told the *Harvard Gazette*. "We treat them with an antibiotic that only allows the most productive cells to survive and make it on to the next round of evolution."

Microbes could generate energy from wastewater and other sources. At UCLA, James Liao was engineering microbes to produce energy and chemicals from algae. A host of companies marketed microbes, including Illumin at the University of Southern California, Emefcy in Israel, and Pilus Energy in Germany and Poland, and others vied to produce their own versions of microbial fuel cells (MFCs). One such cell, called the EcoVolt, was a shipping container–sized reactor that used microbes to generate energy from beer. In Oregon, Widmer Brothers Brewing worried that four gallons of water were required to produce one gallon of beer, and the wastewater was discharged at huge cost. A new fuel cell let the brewer reuse its wastewater, creating energy from microbial reactions. Hong Liu, the company's cofounder and a professor at Oregon State University's ecological engineering department, claimed it could provide a low-cost way of treating wastewater and generating energy for developing nations.

Microbes already treated wastewater in more than a hundred countries around the world in an industry worth some $5 billion annually, featuring such companies as Norweco, BioGP, Clear Blue in Salem, Oregon, and others marketing to wineries, distilleries, breweries, industries, and food processors. Clear Blue had served numerous distilleries, including Real Ale and Blanco Brewery in Texas and Scheid Vineyards and Conn Winery in California. More than two hundred microbial power plants were in use in the United States, with more still in Europe, primarily at dairy farms, where they generated power from manure. Several other companies, for example Microbial Discovery Group and Waste Engineering, explored the use of such microbes as *E. coli, Clostridium*, and *Pseudomonas* to make highly soluble radioactive uranium insoluble in rock. Researchers at the University of Manchester uncovered one microbe that could do so, and those like John Coates at Berkeley had discovered several others.

A new area of business interest was microbiologicals to replace pesticides in agriculture. For years farmers had employed an insect-killing toxin in *Bacillus thuringiensis*, the gene for which was engineered into most cotton and corn grown in the United States. Now several companies, from the giants Bayer, Monsanto, and DuPont to the small, recently started NewLeaf Symbiotics, Indigo Agriculture, and BioConsortia, raced to test thousands of bacteria for their pesticide or biostimulant properties.

Ancient Babylonians brewed beer and Greeks and Romans fine wine, while native Americans made beer from corn for thousands of years. The December 1620 entry in the *Mayflower* journal complained of "our victuals being much spent, especially our beer." Today the microbrew industry in the United States is worth some $19.6 billion, up 22 percent from the year before. New yeast strains are being developed to meet the millennial demand for local brews and spirits, and for the new and organic. The latest fad was special yeasts to promote mental and emotional well-being.

French was getting ready to mop the back in preparation for brewing the newest batch of Free Bird, the local lager. "The steamer blew out last night," she explains. By the time I am done, I am a subscribing customer.

It is easy to think of great scientific advancements happening with big tools costing huge amounts of money on massive scales. Although such advancements are essential to our understanding, sometimes breakthroughs can be found in the smallest of organisms, or home hobbies, right under our noses. Or mouths. Then NASA got involved.

9

Lost City

At the Alkaline Seeps

When the nineteen-year-old industrial chemist Michael Russell worked with toxic chemicals in a London East End factory, he had to measure carefully the ingredients that went into paint primers. In a dirty, cramped backroom, the smallish explorer witnessed firsthand, every day, the significance of nickel in catalyzing reactions at high pressure. At night he went home to listen to his favorite blues recordings and read authors such as Camus and Joyce. Over the years, as it was discovered that much the same conditions could be found surrounding the newly discovered deep-sea vents. Russell, now a geology postdoc working at Ireland's Silvermines deposits, seized on the discoveries to delineate what he thought was life's origin somewhere on the ocean floor. He proposed that microbes originated not at short-lived, acidic undersea volcanoes, however, but at long-lived deep-ocean alkaline vents of seeping catalysts resembling his factory chemicals.

The only trouble was, no one had ever seen such a deep-sea vent. An admirer of Camus's essay "The Myth of Sisyphus," Russell predicted the existence of slowly seeping vents at the ocean's bottom years before the 2003 discovery of just such a strangely beautiful seep in the Atlantic Ocean called Lost City. That ghostly waterworld of stone towers and undulating passageways, created by the reaction of seawater with minerals, was much larger and older than the explosive volcanic black smokers discovered in the 1970s. The discovery of teeming life at ocean-floor volcanic vents helped to convince NASA to search for search for such vents in the ocean-bearing volcanic moons of the outer solar system.

At the same time Russell's Caltech colleague Laurie Barge was studying the electrical charge necessary for life. At Pasadena's Jet Propulsion Lab she built her version of life's chemicals and then coupled them to a power source the size of a car battery. The energy gradients provided a glimpse of life's potential spark. It turned out that alkaline vents generated electricity that life might have harnessed, just as some researchers were using microbes to generate electricity. In one experiment Barge showed how a volcanic vent generates enough power to light a series of light bulbs. In another she recreated the chemistry of seeping vents that Russell and others saw as life's crucible.

Several researchers picked up on the ocean-vent idea, including the University of London biochemist Nick Lane and the Heinrich Heine University botanist Bill Martin. Lane took their arguments to popular videos and books and Martin to sophisticated big-data analysis to identify qualities of the LUCA, pointing to organisms living today at the same cool vents. Then came the discoveries of Earthlike planets by the Kepler probe, including one orbiting Earth's closest star, and a new wave of enthusiasm spread through the field. It seemed the discovery of extraterrestrial life was within our grasp. Perhaps much of life depended not on carbon chemicals, sunlight, and sugars for genetics and metabolism, but on volcanic energy, hydrogen, methane, and carbon dioxide.

NASA's search for life got a boost when researchers at California's alkaline Mono Lake, similar in composition to some of the hot springs of Yellowstone and Kamchatka, claimed that its microbes substituted arsenic for phosphates in its genetic material. That meant life could thrive in some of the most inhospitable planets in the solar system. Life could truly do anything, it seemed. But almost immediately the Mono Lake finding was challenged.

What followed was a science whirlwind, reaching a peak of public interest at precisely the moment the ocean floor was seen to be a potential gold mine of precious minerals and energy. It began with a boy who could not listen in class.

"Not a scientist, never will be"

During World War II, the wiry Michael Russell was doing poorly in school. His teachers railed at the young man who could not seem to find his footing, and who took out his frustration in pranks, truancy, and sports.

Finally he dropped out of school and was assigned a technical job as a factory development "improver" in London's East End. It was a Quaker-owned facility, called Howard's of Ilford, that made paint primers. "Not the cheapest," Russell told me, "and not the best." Still, the workers held shares in the factory, and Russell enjoyed a huge benefit. It would pay so that he could attend night school, and if he went to night school, he received a day off to study. "Improver meant you improved yourself," Russell said. "Like being an apprentice."

His day job was to put toxic chemicals from a coking station into a mix with poisonous phenols that could explode at any time. However, if he added hydrogen to the mix, he got a benign primer that could even be used to make aspirin. His job was to test the different nickel catalysts. "It gave me a sense of the significance of nickel in catalysis and of adding hydrogen to a carbon-bearing molecule," he recalled.

"The irony was. that job gave me a good sense of the conditions that might have led to the origin of life."

Night school was a godsend to a young thinker who had trouble focusing. He graduated with a technical degree and went to the University of Durham for chemistry, his first and true love. Still, he faced a roadblock. When the headmaster read his file, it contained a note from a teacher: "Mr. Russell is not a scientist, never will be, and has no capacity for a university education."

"What's this?"

Russell looked so crushed that the headmaster took pity and sent him with a note to the geology department. There he stumbled into work on volcanoes in the Solomon Islands with a famous Australian geologist, who theorized that many of the world's valuable mineral deposits were the result of ancient deep-sea hot springs. To be sure, German geologists had been suggesting that for many years. The Solomon Islands fieldwork confronted the young scientist with a life-and-death decision one day when his boss wanted to evacuate the entire island based on a rumbling volcano. Taking his readings, Russell determined to override his superior and let people stay. Fortunately, Russell recalled, "I was correct!"

His new interest eventually brought Russell to Canada, then to the Silvermines abandoned quarry in northwest Ireland, an abandoned open-pit zinc mine with strange tiny volcanic mounds. "The luck of being a geologist was that I had worked on mineral deposits and hot springs in Canada and Ireland, where we had found the chimneys," he told me from his cramped office in NASA's Jet Propulsion lab. In Ireland he drew a picture of what he seeking and asked his students if they had ever seen anything like these seeping vents. One of his students replied, "Oh, sure, we have hundreds of those!"

When Russell suggested that these formations were the result of seeping, deep, cool hydrothermal vents, however, almost everyone scoffed. The vent examples found on land were tiny, only a few centimeters in height. At the time Russell's eleven-year old-son Andrew loved playing with the tower-forming "chemical gardens" you could buy out of the back of a comic book. Once, in a fit of pique, the boy smashed his garden in the bathroom. There was a silence. "Hey, Dad!" he called. "These things are hollow!" To Russell, that was a powerful clue. It meant the Silvermines structures were chemical gardens, suggesting that the towering vents known to exist at deep-sea black smokers were not the only possible undersea vents. Gases seeping up from below made the stone chimneys. But the ocean bottom also featured cracks between tectonic plates, where heated water might seep more slowly and steadily, making gentler small, hollow vents. That was how Russell predicted deep-ocean alkaline seeps.

Still, no one believed him. Two papers by Günter Wächtershäuser were indispensable as he moved forward, one in *Science* in 1997, invoking "the pyrite hypothesis," Russell said. Then Wächtershäuser wrote a stunning paper in 2003 on aminations, the precursors to organic molecules. Russell thought, "This is fantastic, now we don't have to worry about the chemistry anymore," and invited the German patent lawyer to speak at their university. Wächtershäuser agreed but admitted he had never spoken in public before. They gave him a date. He never showed up.

At the time Russell's work was gaining steam. But it still faced the roadblock that no such seeping, deep-sea, long-lived alkaline vent had ever been observed.

Lost City

The University of Washington's Deborah Kelley stared down into the sea and climbed into a three-person submarine named Alvin. It sank slowly into pitch darkness, through layers of icy cold and darkness. until the only light came from the bioluminescent fish or Alvin's headlamp. It was 2000, and her team of scientists was working with the National Science Foundation's Ocean Observation Initiative (OSI) on a routine mapping expedition. Suddenly they came across a strange limestone chimney towering 180 feet over the ocean floor. It was the leading edge of something never seen before, a whole new underwater ecosystem about fifteen kilometers from the Atlantic's mid-ocean ridge. It had nothing to do with volcanoes. They returned in 2003 for a detailed study and discovered a remarkable hydrothermal undersea dreamland of spires and mist, fueled by methane and hydrogen. They named it Lost City.

The acres of eerie, undulating plain of dolomite chimneys, smoking sinkholes, and labyrinthine passages extended hundreds of miles along the mid-Atlantic ridge at a point where two continental plates meet. Benign, lasting hundreds of thousands of years—versus the short-lived, hundred-or-so year lifespan of a volcanic vent—and high energy, Lost City water and rock combined to make the beautiful green mineral called olivine, the same mineral used in the United Nations building in Manhattan. But strangely, the Lost City field at first glance looked barren of life, bereft of the giant white tubeworms, fields of clams and mussels, and eyeless giant shrimp that populated the more famous undersea volcanic vents.

Closer examination revealed the reason: the fissures of Lost City chimneys indeed teemed with life—tiny snails, worms, shrimp, and shellfish not found anywhere else in the ocean—but they were hidden in the strange smoking oases that thrived not on sunlight, but on chemical reactions. The larger animals included red stone crabs, grouper-type wreckfish and red-eyed giant eels lacking mouths to eat, relying instead on microbes to digest and excrete their food. Life in Lost City was rich and diverse, and "it made exactly the kind of place where microbial life might have originated," wrote the biochemist Nick Lane.

The reason had to do with the alkaline vents' fascinating geochemistry. Whereas hot volcanic vents pumped out magma at 1200 degrees Fahrenheit into 400-degree water too hot and acidic for life, Lost City produced cool and alkaline water, seeping over hundreds of miles for tens of thousands of years. A trio of researchers, including Russell, Lane, and Bill Martin, seized on the cool vents' geochemical process, called serpentinization, of water hitting minerals as prefiguring the anaerobic metabolisms of methanogens, the Archaea of hot pools, swamps, and our own guts, and acetogens, bacteria like *Clostridium* that live in many of the same habitats. Then it was seen that the geochemistry produced methane, a key ingredient to those organisms, and their theory took further shape.

The vents offered a vital clue, enfolded in a mystery. The answer had everything to do with protons.

Why protons?

It is a great mystery why life uses protons or hydrogen ions, in a quantum mechanism, to make its energy. In many ways, electrons might be much easier to use. Russell and Martin suggested that protons allowed the first cells to convert carbon dioxide and hydrogen, present at Lost City–type vents, into energetic organic molecules. Water produces protons because it consists of a mix of protons and hydroxyl ions joining together and then breaking apart, again and again. Lost City vents produce hydrogen and carbon dioxide, and its small rock pores shelter prototype organic molecules, thus creating energy and transition-metal catalysts shared by many modern cells. Life's origin was thus a channeled response to an energy disequilibrium, with abundant hydrogen, methane, carbon dioxide, mineral catalysts, and microscopic rock pores.

Russell and Martin had the idea that the proton motive allowed the very first protocells to latch onto the geochemical processes already happening at the alkaline seeps, converting gasses and minerals into organic molecules. To do so required a lot of energy, namely the protons available at hot vents. The reduction of carbon dioxide happened with the key molecule used in all things: acetyl coenzyme A (acetyl CoA), the only pathway known that combines carbon dioxide fixation with adenosine triphosphate (ATP) synthesis. ATP is known to every organic chemistry student as the basis of animal metabolism. The amazing thing was that ATP synthase, the machine that generates ATP, *revolves*: the molecule is a tiny rotor spinning in every living thing. A related energy-producing mechanism in plants was a lever.

Protons also powered the waving tiny cellular motors of photosynthesizing chloroplasts and of energy-generating mitochondria. The vents' alkaline water, meeting the acidic seawater, would have produced a natural proton surplus that primitive protocells could tap. "It's like saying the Industrial Revolution started in the U.S. and England because those two countries had so much extra ground coal to burn," Russell recalled. Lost City was a vast electrochemical reactor providing hydrogen, carbon dioxide, mineral catalysts, rock pores, and, most of all, energy. The unusual, shared, ancient metabolisms of Archaea and bacteria may have used exactly those elements to generate power.

Russell sought out the botanist Bill Martin, a tall, funny Texan and former carpenter teaching and researching at Heinrich Heine University in Dusseldorf and writing provocative papers with titles like "The Hydrogen Hypothesis for the First Eukaryote" and, with Michael Russell, "The Rocky Roots of the Acetyl CoA Pathway." As others came to understand the Lost City geochemistry, the two worked out the steps by which rocks and water might become life. Martin empathized with the early, primitive microbes. "I think about these organisms in the deep biosphere," he told me. "They're starved!"

Russell and Martin advanced the metabolism-first idea to the public in popular professional articles and interviews. Martin went further, arguing that complex life began in a fusion billions of years ago when one small anaerobic, methane-eating bacterium took up residence in an anaerobic, methane-creating archaeon. The new tenant provided energy and the landlord the environment, chemistry, and nutrients. With this union, as in a small, growing economy, other units could develop specialties tailored to their talents—respiration, garbage removal, immune defense, and nutrient signaling. As each unit became more efficient at one skill, life developed a control system in a nucleus and grew in size and complexity. The first of those tenants were mitochondria. They did the hard work of metabolism. Biological metabolism thus came as "a direct result of geochemical methane synthesis," Martin told me.

The idea received more support when researchers at Lost City reported finding abiotically produced methane in the vent water. The rock formations spewed mineral-laden water heated from deep inside the Earth. Suddenly the researchers got "lots of press, lots of attention," recalled Martin.

With every news story, the stakes in their debates rose, and as they did, the science disagreements intensified. Lane and Russell diverged over a technical point when Lane's popular writing took off with *The Vital Question: Energy, Evolution, and the Origins of Complex Life* (2015). "I was rather intrigued that he's got a little footnote there which reads, 'I'm sad to say that is now the considered view of Mike Russell too,'" Russell said to me. "The problem is it's awkward to write popular books on a topic that you're an ongoing scientist in."

As they argued, the opposing proponents of the RNA-first world were ignited as well, insisting that genetic material had to precede metabolism. The arguments showed, more than anything else, how close some thought they were getting to the real origin of all life. Fighting over the question of genetics-first or metabolism-first, cell walls–first or cell machinery–first, the field splintered as other teams used more sophisticated, big-data techniques to uncover the most ancient microbial genetics in the tree of life. But what if all three elements—cell walls, metabolism, and the genetic material—originated at the same time?

In Middle Earth

When he was growing up in Merseyside, England, the Cambridge University chemist John Sutherland loved walking along the cold, mud-colored Wirral Peninsula and finding starfish in the shallows. Even at a young age, staring at the tidal pools, he thought endlessly about life's origin. His father was the highly regarded Ian Sutherland, a chemistry professor at Liverpool University, yet the son struggled mightily in mathematics. "Your maths aren't good enough to attend Cambridge," he told John, in one of the father showdowns that can make or break a son's life. But his father also gave him a boost. "Chemistry," he said, "is what you're good at."

The genetics-first researchers had long looked to RNA as the source of life, since it could be both the engine and the control center of the cell. The RNA-first proponents lacked the requisite chemistry to make the correct sugars, to the point that Stanley Miller despaired of ever getting close to life's origin. The first person to provide the chemistry in somewhat convincing form was England's Leslie Orgel, whom I once tracked down to his Salk Institute cubicle on the La Jolla shore. But RNA-first had hit a dead end. That dead end was circumvented in some measure by the papers of Günter Wächtershäuser and the complex chemistry of volcanic vents at the ocean bottom, but the central question remained (to use an automobile metaphor): how could a guidance system, RNA, operate before there was a car engine, metabolism?

At the time Sutherland was an assistant professor in chemistry at Oxford and was looking at the problem from outside the field. "I had a feeling we were framing the question the wrong way," he recalled. "While everyone was bickering about what came first, genetics first or metabolism first or cell walls first, they all agreed *something* came first. What I did was say, hang on, that's an assumption! It's not necessarily true that one or the other must come first!"

Moving to a full professorship in chemistry at Manchester University, Sutherland made the first breakthrough with the doctoral students Matthew Powner and Béatrice Gerland. In 2009 they showed that those genetic elements and early metabolic processes could indeed evolve all at once, together. You could build a compound that was part ribose and, by adding another simple molecule, get a ribonucleotide. After their findings were published in *Nature Chemistry*, the discovery was picked up by the *Guardian*, the *New York Times*, the BBC, and many other media outlets. Sutherland offered "for the first time a plausible explanation . . . for how an information-carrying biological molecule could have emerged through natural processes from chemicals on the primitive Earth," wrote the science reporter

Nicholas Wade in the *New York Times*. Harvard's Jack Szostak described it as "one of the great advances in prebiotic chemistry." The Scripps Institute's Jerry Joyce called it a "near miracle." Nucleotides, the building blocks of RNA, could be formed from simple chemicals in conditions close to those of the early Earth.

In 2015, after Sutherland won a post as a professor at the prestigious Medical Research Council Centre in Cambridge, he showed that you could make nucleic acid precursors simply with hydrogen cyanide, hydrogen sulfide, and ultraviolet light, thought to be present in the early solar system. Those same materials could also make the ingredients for amino acids and lipids for cell walls. "We took hydrogen cyanide from the atmosphere, and used electrons from the reducing power of hydrogen sulfide, and reduced the hydrogen cyanide," Sutherland said to me. "That was the simple thing we discovered. Cyanide is pervasive in the universe. It's in the center of galaxies, gas clouds, protoplanetary disks, and on places like Titan!"

They created some twelve of life's twenty amino acids, but the series of reactions did not happen easily. Rather, each element had to appear separately and sequentially in a central pool of water. Sutherland pictured ancient Earth streams flowing over minerals, bringing the ingredients separately to a pool, bathed with ultraviolet light from an early sun. One had to go back to the early days of the solar system to understand how such conditions might have arisen. What that meant, astonishingly, suggested Wade, was that the giant planets Jupiter and Saturn, which had once been much closer to the sun, accidentally set off a chain of events that gave rise to Earth life. First, the giant planets were driven outward early in the solar system's history, part of a process that unleashed a barrage of asteroids called the Late Heavy Bombardment. In the heat of these impacts 3.8 billion years ago, carbon from the meteorites joined with nitrogen in the Earth's atmosphere to create hydrogen cyanide. It made a poisonous red rain that fell and gathered in pools, reacted with metals, evaporated, and, irradiated by ultraviolet light from the sun, rained again into freshwater pools, where the cyanide reacted with the chemicals again to create and concentrate the precursors of cell walls, cell metabolism, and genetic material, all at one time.

The Sutherland scenario put him in direct opposition to Russell's, Lane's, and Martin's ideas on life's origin at alkaline vents. In 2016 laboratory experiments at the Rensselaer Institute in Troy, New York, modeled a plausible way to make the elements of life at the sea floor in conditions much like those at alkaline seeps. By injecting solutions containing metals into water-filled glass jars, the NASA-funded researcher Linda McGown formed the chimneys observed by Russell. She found the chimney could oligomerize ribonucleotides, a step toward making the building blocks of RNA. "It's a little like Middle Earth in Tolkien's book," McGown said of the strange chemical world she created. There were now two ways that geology led to chemistry, and chemistry to life, but the location where that happened was more intensely disputed with each experiment that was published.

It could have happened many times, in different scenarios. "Life is a sequence of likely events governed by natural laws of chemistry. If you have one hundred billion stars, in one hundred billion galaxies," Sutherland told me, riffing on the Drake equation made famous by Carl Sagan, "could it turn out that life is a universal phenomenon in a certain percent of rocky planets certain distance from a sun?" he asked. "It manifests itself in at least one planet as sentient beings capable of interrogating its presence, which is literally a mind-blowing, dangerous thought!" His idea got a further boost when Curiosity discovered that Mars's Gale Crater had had the very conditions his experiment required—some 4.5 billion years ago.

The only trouble was, Lane and Martin disagreed. "John Sutherland is a good chemist," said Martin. "He is not a biologist."

Sutherland, whose doctorate was in molecular biology, countered that the LUCA Martin was tracking was much farther up the tree of life than Sutherland's protocells. Perhaps the cells had originated in many locations on the early Earth but only survived in the deep sea, which offered protection from the toxic bombardment of meteors and radiation on ancient Earth. LUCA, with its proteins, genes, and cell walls, was nowhere near to life's origin. "Bill has done a fantastic job elucidating what LUCA was," Sutherland said. "That's good, Bill. Stick with that!"

The more sophisticated the experiments became, the more the arguments intensified. As the life seekers seemed to close in on their goal, the ultimate scientific mystery, a missing key remained that of energy, Lane and Russell argued. "You need to bring energy into the system," Martin told me. In Pasadena, the Jet Propulsion Lab researcher Laurie Barge sought to understand the role of energy in life's origin, as a host of other researchers worked to create viable microbial fuel cells.

Power surge

When she was growing up in northern California, Laurie Barge wanted to be an astronomer. As a girl, she read Stephen Hawking's *A Brief History of Time* and thrilled as the NASA probe Voyager shot past Neptune. At Villanova University in Pennsylvania, Barge fought to pursue her dream and graduated with a bachelor of science in astronomy and astrophysics. From there she moved into the University of Southern California's award-winning geology department so that she could be close to the NASA-affiliated research center in Pasadena. To get a taste for industry, she spent a summer helping to search for petroleum for Marathon Oil. She earned her Ph.D. and returned to USC, where, at Pasadena's Jet Propulsion Lab, she built her version of life's early chemicals and then coupled them to a fuel cell in which she replaced high-energy platinum with the iron thought to have been abundant on early Earth. The energy gradients provided a glimpse of life's potential spark.

The cells were similar to batteries in generating electricity and power, but they required fuel, such as hydrogen gas. "We think one important factor was that the planet provides electrical energy at the seafloor," said Barge in a presentation at NASA. "This energy could have kick-started life." She showed how a hydrothermal vent creates enough energy to light a series of bulbs. "Now," she said, "we have a way of testing different materials and environments that could have helped life arise not just on Earth, but possibly on Mars, Europa, and other places in the solar system," she told me.

As a member of the JPL Icy Worlds team at the NASA Astrobiology Institute based at the Ames Research Center in California, Barge created a model hydrothermal vent modeled on Lost City and studied the chimneys that formed in it. "One of the basic functions of life as we know it is the ability to store and use energy," she said. Green rust on Mars was yet another potential site of life, which could have been made when the minuscule levels of oxygen interacted with the primordial soils on the Red Planet. The Earth version might be the green clay of Indonesia's Matano Lake.

As investors raised millions for companies to harness the energy potential of wastewater microbes, power-starved California issued a report that microbes could power hydrothermal fuel cells to provide rural energy. Microbial fuel cells became a hot area of interest that numerous young researchers sought to exploit. The way it worked was this: Bacteria in an anoxic environment attach to a cathode. When they create energy to live, the oxygen waste must go somewhere, so it goes to the cathode, thus creating small amounts of energy. They could be used for wastewater and breweries and municipal treatment plants, but at other places as well. Microbes could harness virtually any biodegradable compound to generate energy. Life was energy, and life's origin seemed closer to being understood.

At the same time the potential mineral riches beneath the sea at black smokers and alkaline vents attracted widespread interest from mining companies. India, Japan, South Korea, and China began their own early stages of exploration. In New Guinea the Canadian company Nautilus Minerals won the world's first undersea mining permit from a national government, seeking gold and silver at the Solwara vents one mile from an active volcano and one and half miles below the ocean's surface. At the ocean's bottom the vents resembled a vision of the world's beginning, with huge colonies of tubeworms, some reddened by the superheated water, so alien and beautiful no words could describe them. The prospect of mining down there jarred residents and activists.

NASA's search for life got a boost when researchers at California's Mono Lake claimed its microbes substituted arsenic for phosphates in metabolism. Life could do anything, it seemed. There was just for one problem: their finding was wrong.

A mistake

Stretched out south of Lake Tahoe in the towering Sierra Mountains of California, Mono Lake flanks Yosemite National Park in a 150-mile basin that collects all the salt of the surrounding valley. An alkaline lake that houses brine shrimp, it is the temporary home for some two million migrant birds every year. Spindly, ghostly towers of carbonate extend their limbs into their air. Here, in a spectacular NASA-sponsored Christmas news conference in 2010, the Lawrence Berkeley National Laboratory researcher Felisa Wolfe-Simon claimed she had found an extreme microbe, the bacterium called GFAJ-1, using arsenic, which is more prevalent on other planets and moons, in the place of phosphorus in its DNA and RNA. "This expands significantly what we think life can do," Wolfe-Simon claimed.

It turned out the research team had mistaken a key chemical. Life in Mono Lake was not using arsenic in place of phosphorus in its DNA and RNA. However, Duquesne University's John Stolz, who played a supporting role in the report, pointed out that "plenty of microbes metabolize arsenic in Mono Lake and in other alkaline lakes." These included the Diamante and La Brava Lakes in Argentina. A hint that arsenic metabolism might have been significant in early evolution was suggested by the discovery of arsenic-enriched layers and arsenic-encrusted carbon globules in the 2.72 billion-year-old stromatolites in Australia's Pilbara.

Some of the Sutherland and Russell claims also ran into problems. Sutherland's chemicals could not be mixed together all at once. They would float away from each other before they could join, so in a pool or a hydrothermal ocean vent the water would dissolve the precursor RNA. Even if nucleic acid was protected in tiny rock crevasses or cracks, and later in a cell wall, its twin functions as information carrier and enzyme were contradictory: An enzyme folds into an intricate shape in order to be highly reactive; a nucleic acid does not change shape and should not react. For his part, Russell was taking heat for coming up with a very specific answer to the multiple sources of organic molecules on early Earth. "His strongest contribution is to understand the complexity of the early Earth's geochemical environment," Robert Hazen told *Nature*. But to focus on alkaline vents, Hazen said, was like starting a game of Twenty Questions by asking, "Is it Winston Churchill, instead of 'is it a man?' It's glorious if you're right, but doesn't help much if you're wrong."

Some went even further. In Florida, the NASA researcher Steven Benner wondered if the whole question was being misframed; he focused his interest in dry-wet cycles on early Mars and the compound boron. "If you have this generational, lingering fundamental disagreement," he told me, "then maybe we are asking the wrong questions." The problem with the Russell and Sutherland formulations was that they broke up over time. They would not produce life, but rather a "tar-like substance," he told the *New York Times*.

Then a 2016 discovery injected a new approach to the argument over life's origin. It came from a young woman in Bill Martin's lab in Germany.

On the Rhine

At a routine lab meeting at Dusseldorf's Heinrich Heine University, the graduate student Madeline Weiss was reporting on the progress of her master's project. Such a report is "not usually a place of major revelation," Martin said and often can be exactly the opposite—a cruel time to admit problems or failure. Weiss was seeking the core genes of the LUCA. She had scanned more than 6.1 million protein-coding genes and removed all the redundancies caused by gene transfer among different organisms. She focused on genes that appeared in two archaeal and two bacterial phyla. "She was doing it without my knowledge," Martin recalled to me. "It was also taking up most of our computer server!"

In the big-data search of some 286,000 protein clusters, she referenced almost every available sequence of every living organism. In so doing, she had narrowed life's core, minimum number of genes to a surprisingly small number: 356. Even more interesting, when she had looked at the organisms that had those genes and what they did, she found they were precisely the acetogens like Clostridia and methanogens like Archaea found at the hydrothermal vents in such places as Lost City.

It was like getting an "electric shock," Martin recalled to me. He sat up. "Whose idea was this?" he asked. It came from the lab bioinformatician, Weiss said. "Okay," he leaned forward. "So, what else did you get?" he asked.

"She was going to tell you," another student chimed in. "But you keep interrupting, so let her give the report!"

The finding, published in *Nature Microbiology* under the beguiling title "The Physiology and Habitat of the Last Universal Common Ancestor," exploded in popular media, so much so that Martin spent the summer of 2016 answering reporters' questions. It suggested, first of all, that life began at cool vents. It also suggested that metabolism came first, and that early life used the local geochemistry to support itself. In fact, early cells were only half alive, relying on the features of their environment and of each other to eke out a hunkered-down version of life.

The claim brought loud objections from such researchers as Sutherland, who argued that the ocean's water would dissolve any preorganic metabolism. It was also suspicious that Weiss's experiment produced exactly the result Martin and Lane predicted. The assumptions in Weiss's big-data approach were wrong, some argued.

"John Sutherland and I really don't have much to say to one another," Martin replied. "The problem is he has to assume very unlikely conditions on the early earth. Our model is much simpler."

In microbiology, the biggest and smallest scales were converging as research intensified, digging into the furthest reaches of the universe and the earliest dawn of Earth with new machine-learning algorithms. The disruptions and arguments percolated into pop culture and even the Vatican, which invited Martin to speak on big data and the origin of life." He showed up in the great Pontifica Scientarium, near where Galileo had been censored five hundred years earlier, only to meet a highly receptive Vatican science committee. "I want to know where things come from," he explained to me. "I want to know specifics, not abstractions. You know, almost everything everyone has said in science is wrong. In two hundred years, scientists will be laughing at many of the things we say today are true."

For his part, Russell was rereading *Moby-Dick* for solace and lingered over Melville's quote, "It is not down in any map, true things never are."

"It won't be tidy," said John Sutherland. "Nothing in history would tell you that."

10

Moonlight

Symbiosis, the Squid, and a New Science

Heat lightning flashed over the ocean. The thirty-nine-year-old zoologist Margaret McFall-Ngai was wading in the Hawaiian surf amidst the torch fishermen, looking for eyes peering in the lapping seawater. Athletic and tall, described by her students as regal, she wore shorts and booties and carried a flashlight and a small net. An avid body surfer with an Irish flair for the dramatic, she had four days to collect enough walnut-sized bobtail squid to bring home to California. McFall-Ngai had no idea whether her project would work. With one of her graduate students, the Iowan Mary Montgomery, she scanned the shallows intently for the tiny squid *Euprymna scolopes* swimming in the water column around Coconut Island, but now the animals were not turning up. The two women had to find out whether the squids could be transported to their University of California lab.

E. scolopes was a beautiful creature, spotted and multicolored, with a cunning survival strategy called bioluminescence. It housed bacteria in a specialized organ near its stomach that made a night-time glow like moonlight, camouflaging it from its monk seal preda-tors as it floated above them in the shallows. During the day the squid carefully tucked itself into ocean sediment. At sunset it awoke and imbibed a fresh set of bacteria, which helped it to grow and thrive and luminesce. In return, the squid fed its bacteria a mix of sugar and amino acids in a special light organ it created expressly for that pur-pose. As the nocturnal squids floated they lit up the water like tiny stars, making themselves invisible.

McFall-Ngai understood what it was to camouflage yourself from those with more power than you. Raised in a strict Catholic home in the small beach town of Del Mar, north of San Diego, and sent to an girls' prep school in Los Angeles, she was a softball pitcher, volleyball player, and basketball forward, but waited before telling her mother she wanted to study biology. At the Jesuit University of San Francisco she was required to study Roman ethics, Greek philosophy, and logic, and she loved decoding the rules of complicated, ordered systems. She preferred studying life yet was so shy she begged, successfully, to be relieved of the undergraduate speech requirement.

In 1990 she was trying to prove that microbes were the keys to understanding animal behavior and health. The prevailing model was that of competition between host and invader. McFall-Ngai was con-vinced that cooperation was even more important to survival. "The arguments of Lynn Margulis were so logical and compelling," McFall-Ngai told me. Lynn Margulis tracked her down to speak at the world's first symbiosis conference, held in Bellagio, Italy, in 1989, bringing together some of the new field's leading researchers. Understanding symbiosis in mammals housing trillions of microbes, McFall-Ngai told the audience, was too complicated for the tools of the time. To prove the importance of symbiosis, she argued, you had to study it in an animal containing a single microorganism.

She rested a second and felt the warm Hawaiian water. A shark fin or shadow of something lurked in the distance. The stones were slippery. A distant bolt of lightning illuminated the surrounding mountains in a garish flash. She spotted a pair of glowing eyes in the water and plunged her net down.

Around her Los Angeles lab, McFall-Ngai was a sophisticated, even glamorous, researcher. The sound of her high heels echoing down the hall was, to her graduate students and postdocs, the sound of authority. But most of all she loved being in the field. Once, after breaking her leg, she gave a presentation and flew back home before going to the hospital. She felt urgently that biology faced a crossroad. People around the world were studying microbes, but most of those people were not talking to each other. In Siena, Italy, Rino Rappuolli was studying vaccines' effects on the microbiosphere. At Sweden's Lund University, Catharina Svanborg and others were looking at pathogens in women's urinary tracts. At Yale, Jo Handelsman was seeking new antibiotics and understandings of antibiotic resistance. It was fine for symbiosis researchers to argue about distant events, like the multibillion-year-old union of a cyanobacterium with a bacterium to make a chloroplast, or of a mitochondrion and bacterium to make a cell's energy factory, but the world was in crisis. Medicine needed quantifiable data to understand the complicated reciprocal signals between microbes and their hosts. The only way to get that was to study a real animal. Not in theory, but in the torchlit waters off Coconut Island, called Moku o Loʻe by the locals, on a night with heat lightning. The bobtail squid went flying into her plastic bag.

California dreaming

Around the time that Margaret McFall-Ngai was a graduate student at UCLA, the deep-sea research submarine Alvin was exploring enormous caches of unknown microbes living around volcanic vents at the ocean bottom in weird symbiotic communities with microbial cousins that fed giant tubeworms and other creatures. They lived also in the hot volcanic pools of Yellowstone alongside the beautiful blue-green and purple mats of cyanobacteria, which we now know were the basis of plants' photosynthetic chloroplasts. It was a revolution in vision. The deep-ocean communities rivaled the tropical rainforests in biomass, depending not on the sun but on the dark energy of volcanic hydrogen sulfide as fuel. Only later did we see some of the same microbes in mangrove swamps in tropics around the world, in sewage effluents and coral reefs, and in our own bodies.

The trouble with the newly discovered microbes was that no one could cultivate them in the lab, at least at first. Then the Indiana scientist Norman Pace took his group to scoop the muck at Yellowstone. The pilgrimages to Yellowstone's Obsidian Pool created a new genomics approach to microbial diversity. The result was bracing— fragments of new organisms, each stranger then the last. Microbe diversity was far greater than anyone had imagined. In fact, many of the early microbes did not exist as specific species at all. For a billion years they traded genes freely, clinging desperately to life in tidal pools and borrowing freely from one another's genes for metabolism, energy, and respiration. The methane-producing Archaea, for instance, often lived with cyanobacteria that did produce oxygen. In a harsh Earth bombarded by meteors, with few continents, vast acidic oceans, and wrenching tides, of course the early microbes banded together to survive.

McFall-Ngai set out to understand the modern version of that world, still dominated by microbes that had outlasted us by billions of years, on which all the plants and animals depended. "A new understanding" was needed, she wrote with her colleagues, "one that recognizes strong interdependencies . . . between these complex multicellular organisms and their associated microbes." Our key activities, things like sleeping, breathing, immune protection, damage repair, and growth, were invented, aided, or controlled by microbes.

The trouble was, no one would listen. To get them to do so, she needed an animal that housed a single bacterium.

"That's what I'm going to do."

Some of evolution's biggest names, such as Stephen Jay Gould and Richard Dawkins, ridiculed the claim that symbiosis drove the significant leaps in evolution. The idea and the outspoken Margulis herself were widely criticized or simply not "discussed in polite circles," as Margulis once observed. She made mistakes. One of the few science organizations that did pay attention was NASA, which put Margulis in charge of its new Planetary Science Division in 1982. The next year the outspoken McFall-Ngai earned her Ph.D. from the University of California, Los Angeles. She was studying the ways luminous ponyfish in the central Philippine Islands used bacterial light to evade predators. It frustrated her that she could not analyze the full duration and intricacy of the symbiotic relationship, because the fish swim so freely in the wild and had such complex behaviors. Years earlier the University of Hawai'i biologist Dick Young had first mentioned to her the bobtail squid. Its single bacterium could live inside or outside the animal, making it much easier to study the effect of symbiosis on both host and invader.

Oh my god, she thought. That's it. After my postdoc, that is what I'm going to do.

McFall-Ngai was already a budding star, named Graduate Woman of the Year at UCLA, but she had not done what she wanted to do most. She read the influential Woese and Fox paper on the gene-sequencing approach to cataloguing microorganisms. "which allowed biologists for the first time to recognize the true diversity, ubiquity and functional capacity of microorganisms," she later wrote. Her postdoctoral work with a leading researcher at the Jules Stein Eye Institute covered human protein biophysics, but then she took a second postdoc at the Scripps Oceanographic Institute in La Jolla, where Dick Young showed her how to catch the bobtail squid. Another postdoc at Scripps, the red-haired, bearded, ponytailed Ned Ruby, studied the same luminous bacterium as it lived freely in the wild. But it could not be grown in the lab, and the Florida-raised Ruby was about to give up.

After 1988 McFall-Ngai came to her faculty position at the University of Southern California, where she asked "every microbiologist she could find" if they would collaborate. No one would do it. She was studying animals and they studied microbes. She ran into Ruby again and asked him to consider assigning a postdoc to work on the bacterium, named *Vibrio fischerii*, *Vibrio* for short. *Vibrio* was a cousin to microbes that caused cholera and gonorrhea, but in the squid it was a benefit. The squid created for the bacterium an elaborate nest in a light organ near its stomach, and *Vibrio* in turn signaled the squid how to grow itself. What could that relationship reveal about the line between sickness and health? He said he would do it for her.

They flew to Hawaii together and set up a small station studying the role of beneficial bacteria in the promotion of health in the squid-*vibrio* model. Some biologists and zoologists were realizing that such symbioses were likely to occur in most animals, including humans, but that these associations occurred as highly complex, multispecies consortia. With support from the National Institutes of Health, the National Science Foundation, and the Office of Naval Research, they began publishing studies showing how much of the squid's early growth was designed to accommodate the bacterium, landing them in *Science* and the *New York Times*. Ten years after she won the Albert S. Raubenheimer Award for Outstanding Junior Faculty at USC in 1994, McFall-Ngai went on to show in 2004 that the bacterial signal to its host could be similar to the toxins produced by *vibrio*'s disease-causing cousins. "That idea opened a door," said Ruby.

Somewhere in the grueling days and nights working together to make a single discovery, McFall-Ngai and Ruby fell in love. She was a big-picture, creative thinker, and he was an expert with detail and a gifted mentor to students. They thrived on brainstorming with each other. "It's a real symbiosis the two of them have," their colleague Nicole Dubilier, at Germany's Max Planck Institute, told *Nature*.

McFall-Ngai discovered, in the early interviews and papers, that she had a knack for communicating science to the public. But still, few people were listening.

Look and look again

As the Carl Woese group in Illinois, in collaboration with the Maryland-based team of the code hunter Craig Venter, analyzed the genome of an extreme microbe discovered at a volcanic vent three kilometers below the Pacific surface, what they found was amazing: it lived in water up to 200 degrees Fahrenheit, under water pressure that equaled the weight of more than 300 Earth atmospheres, and 56 percent of its genes had never been seen before. This was followed by several more extreme-microbe gene sequences, each weirder then the last. The extreme microbes coexist in a symbiotic relationship with the giant red-tipped tubeworms, ghostly fish and crabs, and strange shrimp with eyes on their backs, in complex communities at dozens of turbulent, violent sites on the ocean floor, dark places oozing energy from the middle of the Earth. Analyzing the molecular structure of the tree of life allowed biologists to glimpse the diversity and ubiquity of a hidden world that might well be more common in the galaxy than our visible one.

Two postdocs scooped the muddy, boiling water in a Yellowstone black pool and sequenced the extracted DNA. By this method they found microbes that looked like the most primitive organisms yet. Then they found more unknown species of Archaea. "At first we thought they were only extremophiles," recalled Jamie Foster, McFall-Ngai's graduate student and now a NASA researcher at the University of Florida. "Now we know they live everywhere—in your mouth, your gut, and in the ocean." They shared a slow metabolism unlike anything science had ever seen.

For centuries microbes had helped us make us make wine, beer, and cheese. They were the source of antibiotics. They cleaned up toxic waste. They could take agricultural and municipal waste and create hydrogen for power and fertilizer for crops. A 1990s State of California report suggested that organic waste decomposed by anaerobic microbes could provide up to 10 percent of the state's electrical needs. The idea was spreading that we could turn to microbes to help protect our emotional health and the environment, but it was still a fragmented idea among different people with different agendas, most of whom did not speak to each other. The discoveries were happening, but with little sequence or scientific organization. Somebody needed to put them all together.

The wrong sand

Sometimes they ran into problems. At the start McFall-Ngai flew to Hawaii and Ruby stayed in Los Angeles so he could be there to receive the animals. She sent him a group of squid, and he put them in a tank with beach sand. They died. She sent more that he put in the tank. They died. She tried a third time. They too died in the lab.

"Ned, what is in that tank?" she asked. "Where did you get the sand?"

"Santa Monica Beach."

"They trucked in desert sand! You need real coral sand!"

Once, in later years, the precious cargo got lost on the tarmac in the Minneapolis airport. Another time the animals were delivered to the wrong address. They could not catch enough around Moku o Lo'e and eventually moved to Maunalua Bay just north of Honolulu.

But she loved solving problems. In the Philippines, when she needed to get to a remote mountain fish market, she requisitioned a four-wheel drive Jeep from the university. They told her to fill out a form and wait a week. She walked to the nearest bar and asked, "Who'll drive me to the fish market for ten dollars?" To be closer to her research, in 1996 McFall-Ngai won a position at the University of Hawai'i's Kewalo Marine Laboratory, and Ruby accompanied her. She asked her Los Angeles graduate students, "Who wants to go to Hawaii?"

The gentle, intelligent little animals reproduced faithfully and prolifically, so the researchers made good progress. McFall-Ngai's lab quota of eight to ten pairs could produce 60,000 offspring in a year. First, her team learned that the bacteria signaled the baby squid to modify the organ in its stomach to accommodate the bacterium. In fact, other researchers found examples of bacterial signaling that sent infant marine animals into growth spurts. Without their microbes, some animals never grew up properly.

Next they sequenced, or listed, all the genes of the *Vibrio* bacterium. Until then only the genomes of its three disease-causing cousins had been sequenced. That helped enable the third discovery, in 2004, that the bacterial signal to the animal's immune system was indeed a toxin, much as McFall-Ngai had predicted. A health benefit was signaled in much the same way as a pathology, the difference being a matter of degree or context. Perhaps that was why *Clostridium* can live harmlessly in our bodies until it is triggered to become pathogenic. McFall-Ngai wondered about the differences with the human immune system. She concluded that vertebrate immune systems were designed to "recognize and manage the many complex communities of beneficial microbes" that live inside mammals, and not simply to attack invaders, as medical science had long maintained. She dubbed the idea "care for the community."

Still, few were paying attention. Her graduate student Jamie Foster, who followed her from California to Hawaii, arrived at a conference with a poster announcing their work. "Not one animal biologist came up and talked to me," she recalled.

McFall-Ngai realized it was not enough to publish papers; she would have to bring together many scholars to develop their ideas and make them heard. She applied to the Rockefeller Foundation to create an interdisciplinary conference on symbiosis to meet in Bellagio, Italy, at the same venue where Margulis had launched the field.

She invited about two dozen researchers from different fields from all over the world, including the University of Cambridge's Claire Bryant, who was studying bacterial recognition by mammalian cells; Novartis Italy's Rino Rappuoli, who wondered how antibiotics were reshaping the bacterial world; and the Catholic Health Services doctor Fernando de la Cruz, who studied the ways in which the Industrial Revolution transformed microbial evolution. It was an important meeting, but their excited ideas raced ahead of the technologies. What was required was a systematic coordination of the new analytical tools of the post-genome era, able to process millions of data bits to understand genetic and biochemical signaling of thousands of bacterial species in a mammalian gut. These system analysis tools, with better algorithms, were urgently needed to bring the microbes into our understanding of human biology.

At the time, the National Institutes of Health immunologist Polly Matzinger was proposing a new understanding of germ infection. Until then, medicine followed the 1948 Nobel Prize–winning idea of defeating foreign invaders, called self-nonself. In bold experiments and papers, Matzinger modified the idea to what she called danger-nondanger. It was not the invader, but rather the signal, that mattered to the immune system. Her papers ignited medicine, perhaps because she was a woman and a talented speaker who was not shy about having once worked as a waitress and Playboy bunny. *Business-Week* said she was "standing the immune-system theory on its head." The *New York Times* called her idea a "full-scale challenge to the reigning theory of immunology." But it fit perfectly with what McFall-Ngai had found in the squid.

Using new technologies of high-throughput gene sequencing, researchers were now exploring the many benefits of microbes in the body. "We should view ourselves as a composite of microbial and human cells," wrote the Washington University biologist Jeff Gordon, "in developing a research pipeline for identifying next generation prebiotics and probiotics to prevent or treat metabolic dysfunction and nutritional deficiencies." The Rockefeller Institute publication of McFall-Ngai's symbiosis conference proceedings sold out. The fact that microbes can help shape an animal's growth and development was established, in the squid and then in other organisms, including mice. The twenty or so top scientists in the new field each hit the road, to begin talking in an interdisciplinary way.

A new world

At sunset the nocturnal squid awakens and imbibes its fresh set of bacteria, whose whitish light hides it from predators. But the same bacteria's cousins cause cholera, pertussis, and gonorrhea. The strange, compelling message of microbe-animal symbiosis was that some of the same molecular signals that caused illness could also enhance life.

Commercial and academic interest in the marine microbiome intensified. Berlin-based Cyano Biotech and other companies studied lake and ocean cyanobacteria, the culprits in toxic blooms, as a source of new healthcare and medicinal products. Madrid-based Pharma-Mar sought medicines from deep-sea relatives of streptomycetes. Reykjavik-based Matis sought to understand the microbial diversity of Iceland's hydrothermal vents, while Norwegian scientists studied ocean microbes' potential for degrading oil spills. On the manmade side, the Chicago-based Shedd Aquarium's Marine Microbiome Project made the first global effort to catalog the microbiomes of sea mammals and fish in aquariums. "I think about water differently now," said Chrissy Cabay, Shedd Project Manager. "We live on an aquarium planet."

On land, some companies were making use of terrestrial bioluminescence for cheap lighting and road signage. The nerve-deadening food-poisoning toxin secreted by *Clostridium botulinum* made the key ingredient of high-priced Botox treatments.

The concerted effort of McFall-Ngai and others helped bring the complexities of the microbiome to the center stage of biology and medicine. By the time McFall-Ngai was nearly done teaching in a professorship at the University of Wisconsin, more studies were suggesting that some forms of depression, obesity, and attention-deficit hyperactivity disorder (ADHD) were linked to deficits in our microbes, perhaps caused by the overuse of antibiotics. The human gut microbiome calibrated our burning of calories and milk metabolism, and also served as a "tuning fork," one study showed, of a child's developing immune system. Inflammatory bowel disease and Crohn's disease, both highly dangerous and rising in incidence, were correlated with microbial deficits. Type 1 diabetes was linked to exposure to an *Enterovirus*. According to the Max Planck Institute researcher Kirstin Berer, even multiple sclerosis, a disease in which the gut microbes seem to turn against its nerve cells, might be addressed by a simple transplant of microbes from a healthy person.

Recurrent *Clostridium difficile* infections, inflammatory bowel disease, and Crohn's disease all defied normal antibiotic, radiation, and antiviral treatments. As a last resort, doctors took to fecal transplants from healthy donors, and achieved remarkable results for some patients. A fecal transplant is a procedure by which fecal matter is taken from a tested donor, diluted in saline, strained, and placed in a patient by colonoscopy or enema. The purpose is to replace good bacteria that have been killed or suppressed, usually by antibiotics, allowing pathogens like *Clostridium difficile* to overpopulate the gut. The American Society for Microbiology convened a "Bugs as Drugs" colloquium to devise ways to employ microbes themselves in the fights against cancer and antibiotic resistance. Some of these included microbes to tamp down the immune response, for those gastrointestinal diseases, or pathogenic microbes to ramp it up, as a way of attacking cancer tumors, or to augment radiation or other conventional antibiotic drug treatments.

There were important differences by age, gender, ethnicity, and region in the characteristics of the human gut microbiome. The children of rural Venezuela, for instance, process lactose better than those of America. Some 10 percent of the carbohydrates in a Western diet would be unprocessed if it were not for gut microbes. This helped explain how giving antibiotics to cattle fattens them up. Some research-ers argued that microbes could exacerbate "Asian flush," the fact that some East Asians do not process alcoholic drinks as readily as some Caucasians. A seaweed bacterium improved the ability of Japanese to digest the seaweed's nutrients, even conferring its metabolizing gene to their gut microbes. Antibiotic treatment of Alzheimer's-prone mice, it was shown, could postpone the disease, but only in males. Perhaps the whole concept of pathogenic versus beneficial microbes was flawed, argued the Johns Hopkins researcher Arturo Casa-Devall. Microorganisms adapted to the bodies of animals, which in turn adapted to them, each living and communicating in ways to maximize their survival.

At Washington University Gordon was pursuing groundbreaking research on new ways to understand the human microbiome. For one, it helped starving people recover. In developing nations, microbes can cause malnutrition even in young people who were getting enough to eat and could be responsible when starving people are unable to uti-lize calories when fed. The majority of these microbes live in our gut, belonging to all three domains of life, and a host of animal studies suggested that the gut microbiome affected obesity, diabetes, heart disease, and psychiatric and sleep disorders. Perhaps the most con-troversial claim about microbes concerned the apparent increase in cases of late-onset autism, with which an excess of up to twenty-five different strains of the combative microbe *Clostridium* seemed to be linked.

The need for expanding the international human microbiome proj-ect, on a larger order than the human genome project, was gaining in-stitutional support. The researchers behind the Human Microbiome Project now used the new tools of multiple international teams, from Singapore to Scotland, including metagenomics, big data, the NIH's Roadmap initiative, new highly parallel DNA sequencers, and mass spectrometers to unite medical and environmental microbiology. The goal was to come up with new ways of defining disease and health, even to understand what was natural and what was unnatural. Manipulating our microbiomes, many were coming to feel, could help address multiple looming crises.

A core microbiome

As research for the Human Microbiome Project progressed, the idea that much of animal biology was missing out on the importance of microbes was becoming more widely accepted. Now when Jamie Foster, McFall-Ngai's former graduate student, who had since become an associate professor, spoke at conferences, rooms were packed. McFall-Ngai published an essay in *Nature* suggesting that mammals shared a core microbiome that provided selective advantages, such as enhanced digestion. From her Wisconsin farmhouse, McFall-Ngai was thinking hard about how to promote these ideas to reshape the way biology was taught and medicine was practiced. She and Ruby hosted an annual powwow of squid researchers, most of them graduates of their two labs, to discuss the latest developments in their field. It was an idealistic goal of sharing ideas, amplifying what any one of them might do on their own. Sometimes they all wore squid hats.

One graduate of Ruby's lab, the Loyola University Stritch School of Medicine's Karen Visick, was using the squid to study the rapidly increasing virulence of hospital biofilms. Her group determined that biofilm formation is a key stage in symbiotic colonization by identifying the genes and regulators necessary for polysaccharide production, symbiotic biofilm formation, and host colonization, and she was named a Fellow of the American Society for Microbiology in 2016, as resistant biofilms and strains of staph, MRSA, and other pathogens endemic in hospitals were crippling young and old alike, and doctors were finding that they had a dwindling arsenal with which to fight them. "The *Vibrio* form a biofilm on the squid's light organ," she explained to me, "then wait until enough are around to ignite the gene that produces light." In turn, the squid feeds the microbe a diet that brightens its luminescence.

Now people could see the direct cause-and-effect results in a highly complex relationship, and they paid attention. McFall-Ngai was admitted to the American Academy of Arts and Sciences and the National Academy of Sciences. She served as organizer of the annual meeting of the American Society for Microbiology and applied for and won a Guggenheim Fellowship to understand human microbes and our immune system, and to pull together the main players in the various fields to make a definitive statement on the role of microbes in evolution and development. The understanding that life's diversity was mostly unseen was "vastly transformative," she noted, but still not fully appreciated. For years researchers had focused on pathogenic microbes; now, however, it was seen to be more helpful to focus on health, not on illness. You could not understand new or surging diseases without figuring out the core microbiome of a healthy human being.

She asked such stalwarts as Harvard's Andrew Knoll and California's Sarkis Mazmanian, as well as Thomas Bosch of Christian-Albrechts University and Nicole Dubilier of the Max Planck Institute, both in Germany, to join in the effort to write a sweeping call for a "new imperative for the life sciences," teaching the ubiquity and diversity of symbiosis in the living world. After months and months of wrangling, they produced a strong document. But *Science* editors balked. They wanted to refocus the article on evo-devo, she recalled, which argued that steps in evolution follow steps in our growth and development. They wanted Knoll, a book author and another National Academy of Sciences member, to recast his section altogether. "It was like we were saying something controversial or new," complained McFall-Ngai.

Finally, after eight months of recasting and endless editing, she polled her twenty-five coauthors. With the grant from Guggenheim, she and her coauthors withdrew their accepted article from the journal. It appeared in the smaller but more widely available journal *Proceedings of the National Academy of Sciences of the United States of America.*

The new questions build on the old, they argued. Animals had their origin in the union of two microbes, and symbiosis contributed critically to the evolution and specialization of new species. It allowed organisms to share microbial gut communities in much the same way as the original microbes shared genes. The whole concept of a tree of life mapping separate species was outdated. It was more of a partnership among species and with the environment that shaped the living world. Microbes played a key role in brain development, behavior, the immune system, weight, mood, and even our sleep cycles. What was the difference between pathogenesis and a beneficial association? Pathogens were a normal part of the body. It could be that disease was a matter of "normal microbes going over to the dark side," she said. After focusing so much on pathogenesis, researchers now saw that virulence factors and toxins were part of the normal conversation a microbe had with its host animal. They were present all the time. Disease depended on the context and nuances of the signals.

Life on Earth, and in our bodies, exists as a "series of nested ecosystems," McFall-Ngai wrote, meaning that the communities of microbes on our palms differed from those on our wrists, or in the crook of our elbows. Microbes were involved at each and every level. The free-living microbes affected many biogeochemical cycles and processes. They seeded the clouds for rain and cooled the Earth when it became overheated. "I look around the surrounding farms and wonder why mile after mile is planted with one crop, feed corn," McFall-Ngai said to me after shoveling snow in her Wisconsin farmhouse. "I put in a recent article that every animal is shaped by the microbial world. A reviewer claimed I couldn't really say that!" She laughed. "I told the editor, 'I'd like the reviewer to name me one that is not!'"

She inspired a new generation of women to address symbiosis, such as Caltech's Dianne Newman, who studied lung biofilms and respiratory disease. "She's one of my heroes, an animal biologist who came to be one of the greatest spokeswomen for microbes you can imagine," said Newman, who also teaches Massachusetts's prestigious annual Woods Hole Oceanographic Institute summer course on microbial ecology. The first assignment Newman gave her students there was to ignite swamp gas, created by the same methane-producing microbes studied by Woese and Pace in Yellowstone, and present in our gut when we pass wind, and possibly the source of the methane detection on Mars by the Curiosity rover. McFall-Ngai was asked to coauthor the National Research Council white paper "A New Biology for the 21st Century."

A host of young scientists began using social media to write about the new research. I followed them on Twitter as they attended meetings, questioned presenters, and shared new papers. A prominent science journalist for the *Atlantic* and *Scientific American*, Ed Yong, wrote a profile of McFall-Ngai in *Nature*. The ideas were catching on. Then the Pacific Biosciences Research Center in Honolulu called.

Surf under sky

The highly regarded Hawaiian Pacific Biosciences Research Center, which is in charge of Kewalo National Laboratory, needed a new director. Its high-powered list of National Institutes of Health and National Science Foundation researchers included some of the best biomedical, genetic, and pharmaceutical researchers studying marine animals as holders of clues to multiple medical crises. The Center asked McFall-Ngai to give a speech in March of that year.

By then the small squid had become a widely known model organism in one of the fastest-growing fields of bioscience. So intelligent that it squirted ink that took its own shape, it had a huge brain for its body size. McFall-Ngai's enthusiasm and sense of humor about the squid had helped bring animal-microbe symbiosis to a peak of interest. "She doesn't deal in little ideas," her friend the Cornell entomologist Angela Douglas told *Nature*. "She challenged conventional thinking about the immune system," said New York University's Langone Medical Center Director, David Sabatini, "with profound implications for medicine."

The room at the Pacific Biosciences Research Center was packed by the time McFall-Ngai got up to speak. "Biology is at an inflection point," she said. "Understanding the role of microbes in earth climate, life evolution, and human health is the biggest revolution in the field since Darwin."

Still, the origin of the idea of symbiosis remained unaccountably controversial. A prominent biologist lectured for an hour on symbiosis and did not mention Lynn Margulis once, McFall-Ngai recalled. "Dianne Newman called me, because we taught this Bio 101 course together," said McFall-Ngai. "She heard one of her colleagues say that Lynn Margulis was a charlatan. I said Margulis was the key spokesperson at the beginning of this field. Very few scientists will be remembered after they are gone. She will be."

It was nevertheless clear that much of life was based not on competition, but on mutualism, and the overuse of antibiotics could be a part of the rise of obesity, diabetes, asthma, and other chronic conditions. Introducing a stronger antibiotic may eradicate a disease, but doing so also leaves behind only the strongest, most resistant bacteria.

What excited people other than biologists was the vast genetic capability of microbes that could consume sulfur, petroleum, arsenic, and heavy metals, sequester radioactive materials, and survive in many parts of the human body. A University of Duluth researcher, Jessica Sieber, studied the human gut symbiosis of methane makers with healthful butyrate-producing, or short chain fatty acid (SFCA)–creating, bacteria. The ability of microbes to breathe gases other than oxygen was important. Their potential as water remediators and new energy sources attracted greater interest.

The Pacific Center called McFall-Ngai back. They wanted her as director and Ruby as senior researcher. Over the next year she poured her time and resources into revitalizing the world's leading marine research facility. The idea that people had ignored or disparaged the potential of microbes to help guard our health had arrived.

At the crowded Boston World Trade Center, McFall-Ngai raced from session to session of the American Society for Microbiology's 2016 meeting. She sat in the back in her jeans, black sweater, and pink-and-blue scarf, leaning forward intently, greeting friends and consulting the list of conference speakers she had meticulously scheduled on a slip of paper. The "new biology" mantra that much of life is commensal, not through altruism but through the sheer drive to survive, was suddenly being applied to numerous problems. With the rates of antibiotic resistance and climate change reaching levels of global crisis, the U.S. and European governments offered financial incentives to companies to find new classes of antibiotics, sources of energy, or ways to clean up the carbon we poured into the air. "The best pathologists in the world are studying the microbiome," she told me. "They understand!" Several companies took up the call to capitalize on the discoveries. The question was how fast one could make use of microbes that had taken billions of years to shape the planet. Time was, it could be said, running short.

PART THREE

Microbes and Money

A Sustainable Future

11

A Universe Within

The New Microbial Medicine

A young northern Virginia couple painstakingly checks the glucose monitor of their type 1 diabetic toddler when going out on a Saturday night, watching his levels by the minute by measuring every raisin he eats and every glass of juice he drinks. Life for them is a nightmare of monitoring, vigilance, and exhaustive calorie counting. The father worries his previous bout with an *Enterovirus* triggered the autoimmune response that caused the boy's condition. But could a tiny virus have triggered a lifelong condition? Microbes help tune the health of the human body, from the anaerobes in the gut to the staph and strep in our mouths and nasal passages, and some of them may be triggering the mysterious increases in incidences of diseases like type 1 diabetes and depression. By the 2010s medical researchers were racing to understand how microbes shaped our lives in profound, hidden ways.

This rapid improvement in our knowledge of the roles microbes play in preserving health and fighting disease happened as new maladies reached epidemic proportions. Obesity was up from 12 percent to 30 percent in America since 1990. The incidence of type 1 diabetes was doubling every generation, and rates of asthma, food allergies, insomnia, and depression were increasing. Hospitals were combating a dozen strains of highly resistant biofilm-forming bacteria. We were refighting the battle against infectious diseases we thought we had won. A host of new studies, from the National Institutes of Health International Human Microbiome Project to the Alfred P. Sloan Foundation's Home Microbiome Study, along with several international microbiome initiatives, all suggested that disruptions in the natural microbiome may play a key role in at least some of the world's health crises.

Our body's natural bacteria digest lactose, helped process vitamins, protect the vagina, prevent obesity, and monitor asthma and diabetes. They break down food, produce anti-inflammatory chemicals, and train the immune system to distinguish friend from foe. They number some one to four hundred trillion cells, mostly bacteria, including also Archaea, fungi, and an almost uncountable number of viruses, otherwise known as bad news wrapped in protein. When we move, we carry them from the old apartment to the new home. If the detection technologies improve, they can help to identify criminals. They colonize newly built hospitals in a predictable pattern. The most recent work on microbes and modern health suggests that some half dozen phyla may offer potential solutions to the problems of antibiotic resistance. From the four pathogens of the respiratory system to the *Helicobacter pylori* in the stomach, some microbes previously considered to be toxic may help us to metabolize nutrients, prevent infection, and even sleep better.

By 2016 some two million Americans a year were contracting antibiotic-resistant infections, often in hospitals and doctors' and dentists' offices, and some 23,000 a year were dying from them. In New York, Peggy Lillis went in for a root canal and four days later was dead. In Texas, eight-year-old Nick Johnson contracted MRSA and nearly died. After a hospitalization in India, a Nevada woman died in August 2016 of a carbapenem-resistant infection that defied all twenty-six American antibiotics. Tom Fried, director of the Centers for Disease Control, called carbapenem-resistant Enterobacteriaceae (CRE) a "nightmare bacteria," killing half the people it infected. *Clostridium difficile* infected 15,000 Americans a year. The danger was that bacteria were smarter than we were, threatening to turn back the clock on a century of medical advances.

A host of researchers sought to adapt next-generation DNA sequencers and sophisticated data processers and algorithms. These, along with an explosion of studies in remote regions on promising animal models, would help us to understand the microbes in the body, the home, the hospital, and the field. These researchers sought to promote a paradigm shift in the understanding of our place in the hidden microbe-human interface, as well as a hawker's bazaar of competing claims. For young children facing a life of careful parsing of every bite of food, the task became one of separating fact from fiction.

It began with ulcers.

Friends with benefits

In 1984, in a darkened university lab in Perth, Australia, Barry Marshall regarded the jar in the fluorescent light, then quickly gulped down his infectious mixture. The story of his hunch about a gut bacterium, *Helicobacter pylori*, was one of a medical miracle as well as a 1993 Nobel Prize. When Marshall swallowed his own brew, he gave himself a stomach infection, showing that stomach ulcers were the results of microbes in the wall of the stomach. The discovery led to an all-out antibiotic assault. "The only good *Helicobacter pylori* is a dead *Helicobacter pylori*," concluded a well-known Baylor College of Medicine gastroenterologist writing in the *Lancet* in 1997. The promise of the discovery revolutionized our understanding of one illness, producing a media and business bonanza as well as an unexpected result.

The result was fewer ulcers, but also an obesity epidemic. Around the same time the medical doctor Martin Blaser was the salmonella surveillance officer for the Atlanta office of the Centers for Disease Control and Prevention. Traveling over country roads to investigate outbreaks on rural farms, Blaser witnessed firsthand the huge amounts of antibiotics that farmers were adding to their feed. The antibiotics fattened the cattle faster, and more cheaply, on less food than required by a medicineless diet. Blaser and his colleagues spent years studying the effects of antibiotics on weight by experimenting with baby mice, discovering that a combination of fatty foods and multiple antibiotics at a young age increased the rates of adult obesity. Through epidemiologic studies, Blaser and others linked antibiotics with the risk of asthma, food allergies, celiac disease, ulcerative colitis, eczema, childhood diabetes, hay fever, esophageal reflux and cancer, Crohn's disease, and even autism in humans. Past plagues struck quickly, but these modern plagues destroyed quality of life over long periods of time.

By 2016 *Helicobacter pylori* offered a prime example of microbes' opposing effects in the body. After we learned that the bacterium, present in our ancestors "since well before we became humans," according to Blaser, now the chief of pathology at the New York University School of Medicine, caused ulcers, we made it the target of multiple successful drugs. Today *H. pylori* is found in only about 10 percent of American children. Quenching the bacterium's inflammatory effects with antibiotics reduced gastric reflux and esophageal cancer. Chronic diseases, however, such as asthma and obesity were associated with the loss of *H. pylori*. An old foe was, under the right conditions, also an important friend.

Blaser went on to become the George and Muriel Singer Professor of Medicine, and the Director of the Human Microbiome Program at the NYU School of Medicine and a cofounder of the *Bellevue Liter-*

ary Review. He engineered the first tests for identifying anti–*H. pylori* antibodies in the blood and became famous for proving its beneficial effects. But it was Blaser's graduate student, the Silicon Valley–raised Californian Laurie Cox, who completed two projects providing suggestive data to prove his theory.

Cox did her first experiment with San Francisco Bay mud at age five, sticking it into a two-liter soda bottle and watching the microbes turn the mud magenta, then mauve, then green, then blue. "It was a Winogradsky column," she told me, "though I wouldn't have known it!" After working as a technician for a company owned by her father, Mike Cox, called Anaerobe, Cox came to Blaser's group at NYU, where she exposed mice to low doses of penicillin at two separate times — as newborns and when they were four weeks old. Twenty weeks in, the mice that received penicillin from birth were heavier and fatter, especially the males. These mice had very different gut microbes from the controlled mice. A high-fat diet only amplified this effect, especially in females. "The key," she told me, "was that the response was delayed. Their gut microbes went back to normal [after the four weeks], but years later, in adulthood, those mice became obese."

In the next experiment, Cox used short bursts of medicine and found once again that only four weeks of early exposure changed the mice's gut microbes. The gut communities returned to normal after eight weeks, yet the mice still gained weight, and their immune systems weakened. Concluding that there was a critical window when microbial disruption could affect the body for long periods of time, she transplanted the microbes from the antibiotic to the germ-free mice. These recipients also gained weight without any exposure to the medicine. "That proved cause and effect," she said. "We showed that gut microbes help determine the ways the body processes food."

One mystery of the global obesity epidemic was that it came so suddenly, over a mere four decades. Blaser and Cox believed antibiotics disrupted the diverse, eons-old ecology of human microbes that evolved with our species, particularly very early in our lives. "The microbes that constitute your microbiome are generally acquired early in life, surprisingly, by the age of three," Blaser wrote.

The same held true for *Helicobacter*. Blaser believed that in an individual's younger years, the bacterium aided in regulating stomach acid. Without this function, we are at a heightened risk for esophageal cancer. *H. pylori* also muted hormonal signals such as ghrelin, a chemical that alerts the brain that it is hungry when the stomach is empty. To make things even more complicated, the microbe was only beneficial to a human younger than forty; therefore, ideally Blaser would like to immunize children with the bacteria at a young age, then exterminate in later in life when the bacteria begins to cause problems. Infants were the most vulnerable. The disappearance of natural gut bacteria at this stage of development was a key factor in later obesity. Both mouse and human children who received *Helicobacter*-killing antibiotics within the first six months of their lives were more likely to be fatter at the age of seven than children who did not have exposure in their first six months. "Our microbiome keeps you healthy and parts of it are disappearing," he concluded. "*Helicobacter pylori* is now an endangered species."

One solution was to reduce prescriptions for antibiotics in children. "We need narrow-spectrum antibiotics designed to knock out the pathogenic bacteria without disrupting the health-promoting ones. This will make it possible to treat serious infections with less collateral effect," Blaser told the FDA. The other was to stop giving antibiotics to feed animals to fatten them faster.

As Cox moved onto a postdoctoral post at Brigham and Women's Hospital in Boston, she turned to studying how gut microbes affect the brain. She was involved in a study on the inflammation-inducing microbes found in multiple sclerosis patients, who, like all of us, faced a crisis in the antibiotic resistance of old pathogens. In 2016 she found that MS patients had altered gut microbiomes, but those receiving treatment saw a reversal of the alterations. "Microbes could be driving some of the immune changes we see in MS," she said. The same held for Parkinson's. There was even a suggestive study on Alzheimer's in which an antibiotic regimen seemed to prevent its onset in male mice. A search was on for new antibiotics. "It really is a new frontier," Cox concluded.

The task was formidable. One place people looked for new drugs against the rapidly escalating epidemic of resistance was in an old target—dirt—but now with new tools of molecular analysis.

"We are in a totally different universe"

From a rolling farm in Maine, American and German researchers with two companies were sifting through soil using a new, hand-held remote tool that cultured new microbes in their own environments. Analyzing the soil back in the lab, they came across a bacterium that produced a new antibiotic they dubbed teixobactin, which proved effective against bacteria that resisted all other antibiotics. The Northeastern University professors Slava Epstein and Kim Lewis created the company Novobiotic to develop teixobactin into one of the first new antibiotics in twenty years, specifically targeting MRSA and *Clostridium difficile*, the highly dangerous and contagious cause of colitis.

The researchers at Boston's Northeastern University and Vancouver's University of British Columbia found the compound, which inhibits cell wall synthesis, by culturing a wild microorganism in the lab—a rare feat. "It's a black art," the McMaster University researcher Gerry Wright told the *Canadian Medical Association Journal* about the requirements for culturing new bacteria found in the wild. They had solicited a diversity of soil samples from friends and their families all over the world. They found teixobactin several times, in several different places. "It has astounding potency," Epstein told me. The bacterium's genome was then sequenced by new high-throughput machines and by the new tools of proteomics, the functions of either of which would have taken a whole career in the golden age of antibiotic research in the 1950s. Epstein said of the techniques, "We are now in a totally different universe."

The key innovation in their work was that formerly some 99 percent of wild bacteria, the sources of new antibiotics, could not be grown in the laboratory. What Lewis and Epstein did was to develop a new tool called the isolation chip, or iChip, basically a tiny board with holes in which they inserted the wild soil, shaking it in water to release the microbes, diluting the sample, mixing it with liquid agar or food, and then dropping the liquid, microbe-rich agar into the iChip. The dilution meant that each hole contained only one kind of microbe, which could still function and eat. Sealing each hole with solid agar, they dunked the iChip into the original soil, allowing the microbes to thrive in their native environments and thus giving "access to things we haven't seen before," Lewis, the team leader, told *Nature.*

Epstein, a bearded, garrulous Soviet émigré, was only beginning. He planned trips to Greenland and elsewhere to characterize the microbes of unexplored regions. "Copernicus, Galileo, Einstein, these were one-man revolutions," he said to me as he sat in front of a whiteboard in his company office. "This is hundreds of women and men, dozens of institutions. There is not a single eureka moment. It is a concerted effort among academia and industry. The revolution is our understanding that 99 percent of microbial diversity is unexplored. It's a discovery of a New World!"

A range of other probiotics now ramped up their claims. Many hoped these compounds—microbes from the deep reaches of Bulgaria, New Zealand, Sicily and elsewhere—could help replace conventional medicines. Italian researchers discovered that a simple probiotic lozenge could reduce the incidence of strep throat in patients. A Bulgarian strain of the much-heralded gut microbe *Lactobacillus* was shown to kill *E. coli* and the strains that cause salmonella. Many of the common microbes in yogurt, sauerkraut, and pickles produced the useful lactic acid that killed some harmful bacteria, even if it could also make our muscles sore.

The probiotics offered a tempting target for consumers looking for natural cures in a human body that, it was seen, depended on its microbiome. The most promising of the probiotics was called VSL-3, sold as a treatment for irritable bowel syndrome. The vast majority of other claims, however, have never been proven, and the nutritional supplements quite often do not even contain the microbes they claim to offer. It was an unregulated industry.

The Bulgarian microbes, for instance, were a key ingredient in the popular East Coast juice smoothies, whose parent company was a major investor in the product. Probiotics like Goodbelly and Culturelle became major sellers, assisting sufferers from scarring and irritable bowel syndrome. The FDA, however, was not convinced.

Other companies rushed into the gap, seeking clinical studies of gut microbe treatments. The companies included Boston's well-known pharmaceutical maker Novartis, Flagship Ventures, Evelo Therapeutics, Elpida, and Johnson and Johnson in the United Kingdom and the United States, as well as GlaxoSmithKline (GSK) in Cambridge and Second Genome in California. A New York company, funded by a brand-new incubator in Harlem called Biospace and written up in the *Atlantic*, sought a probiotic pill to defeat *C. difficile.* "It's a rapidly growing, fast-moving field," said Laurie Cox, who turned her attention to the role of microbes in diseases of old age, even Alzheimer's, as well as Crohn's disease and colitis, where microbes are given to promote regulatory T cells and suppress the immune system.

As of 2016, some 339 clinical trials were under way for testing various microbial treatments, mainly for gastrointestinal diseases in which the microbes would tamp down the immune response. Some 238 of these were for fecal transplants. Of the remaining treatments and companies, the most promising early-stage results were for a completely different target, cancer, where the idea was to give microbes that trigger the immune system. That was the goal at Boston's Evelo Therapeutics; the only problem was that such bacteria were on the border of being pathogens.

The number of articles, websites, and blogs soared, as did investment and business conferences on the future of microbiotech. More questionable claims also piled up, including the one that said microbes could prevent dementia. The popular television doctor Mehmet Oz promoted probiotics for good health. He was brought before Congress to testify on his promotion of diet products, however, such as green coffee beans. For a time I became such a devotee of yogurt, sauerkraut, and pickles that, in the midst of an interview, I had to run out to the bathroom. People broadcast the analysis of their stools. A "Lick Hiker" trekked the United States, licking unclean surfaces from toilet seats to park benches. A probiotic manufacturer promised to restore customers' libidos. Such serious science disappointments followed that the University of California microbiologist Jonathan Eisen started a blog on microbiome hype. Some enthusiasts overstated the effects of gut microbes in preventing Alzheimer's, and many experts continued to overstate the number of microbes in the body. Some thought life could replace phosphorus with arsenic in its DNA and RNA; they were wrong.

Then it was seen that several people around the world had done a better job than we had in keeping their natural biomes intact, so much so that the genes of the gut microbes made for a second genetic makeup in our bodies, and a new wave of interest flowed in.

The second genome

Jeffrey Gordon, at Washington University, was a prominent researcher into the genomic and metabolic foundations of our relationships with gut microbes. Using advanced sequencing machines, Gordon's team analyzed massive data sets he helped to create. By exploring the gut microbiome, he broached the new perspective that humans are an ecosystem made up of symbiotic human and microbial parts, which he described to the Association of American Medical Colleges as "the second genome."

Gordon completed his M.D. at the University of Chicago's Pritzker School of Medicine. His education continued with a residency in medicine at Barnes Hospital in St. Louis, a postdoctoral fellowship in biochemistry and molecular biology at the NIH, and another in gastroenterology at Washington University. For more than thirty years he taught at Washington University in the St. Louis School of Medicine, serving as director of the university's Center for Genome Sciences and Systems Biology and developing powerful new computational approaches to characterize the dynamic operations of human microbiomes. He sought a microbial view of human development, including unlocking the ways in which maturation of the gut microbiota is related to the healthy growth of infants and children, with the goal of ushering in a new era of microbiota-directed therapeutics. By studying twins of different ages, locations, and cultures, taking their microbial communities, and placing them in sterile mice, Gordon's team brought in new evidence about the ways the microbiome affects our health, weight, and metabolism.

In one experiment Gordon and his team transplanted the gut microbes from both fat and thin mice into adult mice that had had no previous exposure to any bacteria (known as gnotobiotic mice). The adult mice that received microbial residue gained more fat than the adult mice that were exposed to the thin mice's microbial content, even though all mice were fed the exact same diet. This data had a major impact on human nutrition in that the nutrient and caloric value of foods were perhaps not absolute, but instead influenced by the gut microbiomes of consumers.

The most surprising aspect of this series of studies was an experiment showing that microbes from a thin twin can take over the gut of a mouse previously exposed to microbes of a fat twin. Gordon called this result the "battle of the microbiota," where gut microbes from thin twin mice were placed in the same cage as gut bacteria from obese twins. The mice in the cage ended up sharing bacteria from the thin twin's microbial community, which "resulted in weight loss and a correction of the metabolic abnormalities." With the right diet, the mouse lost weight. However, the opposite has not shown the same results. No matter what the mouse ingests, microbial content from a fat mouse cannot overcome the gut of a thin mouse.

Furthering the research, Gordon's former colleague Vanessa

Ridaura determined from a national survey the healthiest and un-healthiest diets. Using this food with mice, the researchers repeated the experiment, putting slender and fat mice together in a cage and giving them either the healthy or unhealthy feed. The fat mice that received fatty food kept their microbial content and stayed fat. The "thin" microbial content prevailed only if the mice ate the healthy pellets. It seems that, when properly nourished, gut bacteria can better sustain themselves by picking up the "good" bacteria.

It appeared that an individual's second genome might create as large an impact on our health as the genes we inherit from our family. One major difference, however, was that inherited genes were generally stable, while we may be able to alter and improve the second genome. "We are assimilated by the microbes that dominate the planet. Like every other animal, we have to adapt, coexist, and benefit from the microbial world," Gordon told reporters. As he moved from mouse studies to human studies, he urged caution. It was still unclear which bacteria were responsible for which effects. Gordon's hope was to create pure mixtures of bacteria and treat the failing microbiome with "synbiotics," a sustainable and more powerful probiotic that, ideally, would be administered along with prebiotic nutrients.

At the same time, the issue of cesarean births became another area of interest. Babies who passed through the mother's birth canal picked up her microbes, as they did when they breast-fed. Laurie Cox and others went on to correlate rates of obesity in British children with cesarean births, but it was not a one-to-one relationship. "There are a lot of factors in human obesity—diet, lifestyle, genetics, and the microbiome," she told me when I pointed out that my children had come into the world by cesarean birth and were fit. "We found a slightly elevated risk of obesity in those children."

Partly for that reason, findings from a University of Michigan study showed no correlation with specific human gut microbes and obesity. Comparing some ten studies, researchers saw no carryover or reproducibility of results. Part of the trouble was that other risk factors could not be controlled in humans as easily as in mice. The new microbial frontier could be compared to the difficulties of managing any ecosystem. When the wolves were culled from Yellowstone National Park for instance, deer proliferated and damaged the park's trees. When wolves were reintroduced, the deer population plummeted, and trees returned to the valleys where they had been stripped.

The new perspective, it could be said, was that humans were a traveling ecosystem made up of interlocking, competing, and symbiotic human and microbial parts. Many human-based microbes could be friends in some situations and deadly foes in others. The trick was to manage them. Numerous companies and investors were anxious to capitalize on the discoveries and do just that. The company Second Genome, which can sequence your gut microbiome's genes, signed agreements with Kings College, London. to study the microbiome's effect on food allergies, with University College, Cork, to study its effect on emotional illness, and with Monsanto to use the microbiome technology to accelerate protein discovery. A microbiome bonanza was on the verge of breaking open.

At the same time, antibiotic resistance was reaching a level of worldwide crisis, from hospitals, nursing homes, and doctor's offices to schools, gymnasiums, and daily life.

In the emergency room

A gym rat in California contracts MRSA from a dirty weight bench. A tourist in Key West becomes infected from a scratch by sitting on the curb. A high school athlete in Chicago gets an infection on her leg. Within hours, the wound is red and swollen. She runs a high fever and is rushed to the hospital. Within days, she is fighting for her life.

The problem of antibiotic resistance in hospitals, nursing homes, and our own homes had reached such epic proportions that in June 2015 the Obama Administration hosted a national intervention alert in Washington, D.C. Ana Swanson, of the *Washington Post*, wrote that antibiotic-resistant infections caused at least two million illnesses and

23,000 deaths in the United States each year — more than drug over-doses, cars, or firearms. Sicknesses and deaths caused by antibiotic-related superbugs such as MRSA and *Clostridium difficile* colitis were becoming common. Antibiotic-resistant bacteria threaten not just the very sick, but those undergoing such procedures as joint replace-ments, C-sections, organ transplants, or chemotherapy, as well as patients with diabetes, asthma, and rheumatoid arthritis.

A U.K. government report predicted that by 2050 antimicrobial-resistant infections could kill ten million people a year globally, more than die each year of cancer. The U.S. Food and Drug Administration, the American Public Health Association, the World Health Organiza-tion, and the American Medical Association are all pushing for a ban on the use of antibiotics in animals raised for human consumption.

As a result, the United States, Germany, France, Great Britain, and other countries promoted an anti-antibiotics initiative pinpointing doctors, livestock companies, and drug manufacturers to stop push-ing the use of antibiotics in feed. The chicken giants Tyson and Per-due immediately announced programs to do just that. By the fall of 2015 antibiotic resistance was a top funding focus of the Centers for Disease Control and the National Institutes of Health. The National Science Foundation announced a new grant program to confront the problem.

One key data analyst in the new field was the microbial ecolo-gist Jack Gilbert. A group leader in microbial ecology at the Argonne National Laboratory in Illinois and a professor of surgery at the Uni-versity of Chicago, he was also the director of the Microbiome Center, which leads the microbiome research across the Argonne, the Univer-sity of Chicago, and the Marine Biological Laboratory in Woods Hole. Gilbert managed the Earth Microbiome Project (EMP), an ongoing effort to investigate the microbial diversity of Earth. Gilbert also founded and designed the Home Microbiome Study and the Hospital Microbiome Project, which explore the ways humans interact with bacteria in their homes, in hospitals, and in public spaces. Gilbert also studied the microorganisms in the bodies of critically ill hospital patients, hoping to understand how microbiomes can be both an indicator of health problems and a potential solution. "We want to find ways to treat the microbiome alongside the patient, to make sure both stay healthy," he said to me.

The challenge was to understand microbial growth mechanisms more precisely. For instance, the EMP analyzed microbial communities in their geochemical environments around the world. Launched in 2010 with the goal of analyzing 200,000 samples from global biomes using genomic sequencing to characterize the diversity of microbial life, it produced a global microbiome atlas that includes comprehensive genetic data on each biome. The EMP worked with more than 600 collaborators worldwide to characterize the bacterial assemblages in more than 50,000 ecosystems with an online portal so that anyone may access all the data.

Dangerous biofilms can form in newly built hospitals in a matter of days. In infected nursing homes and hospitals, the task of cleaning was becoming overwhelming. Microbes use horizontal gene transfer in a rapid-deployment mode to become resistant to common cleaners and solvents.

Horizontal gene transfer usually happens in one of three ways: transformation (picking up DNA directly from another organism), transduction (picking up genetic material in the cytoplasm and transmitting it through viruses called phages), and conjugation (bacterial sex). Bacteria can also evolve quickly by sharing transposons, otherwise known as jumping genes. Once the genes are exchanged, antibiotic resistance is conferred via alteration of the permeability of the pathogen membranes or using energy from ATP to remove antibiotics from the cell, via changes in the structure of the antibiotic's target so that it cannot recognize the bacteria, or via use of the bacteria's inherent capabilities to acquire resistance.

That was the state of bacteria. Another huge mystery was why their close cousins, Archaea, never became pathogens in humans. However, some researchers noted that in some multiple sclerosis patients, a form of Archaea increased in the human gut.

Young people who suffer from type 1 diabetes were painfully aware that a viral infection might directly lead to a lifelong chronic illness. Yet some viruses could help animals fight infection. The mysterious, challenging relationship of the microbial world to that of humans and animals was becoming more and more complicated. The vast, nonliving universe of viruses was more explosive than imagined, especially when one of them infected you.

Judicial error

When the thirty-nine-year-old Virginia stay-at-home dad Doug Wood was a toddler in New Zealand, his grandfather took him to the famous Lake Taupo where he fished for the enormous rainbow trout that lived there. As an adult living in northern Virginia, Wood had to quit his law practice to care for his three-year-old type 1 diabetic son while his wife Becky worked as a high-profile law partner in Washington.

It all began one October day when Wood suffered a crushing headache. It grew so bad he was rushed to the emergency room, fearing he had meningitis. After a battery of tests, he was diagnosed with viral meningitis triggered by an enterovirus, a relative of the virus that causes the common cold. In children it causes coldlike symptoms. In adults it can sometimes cause crushing headaches, inflammation, eye pain, and extreme sensitivity to light and noise.

Wood's son Ben got the virus, and four weeks later he was diagnosed with type 1 (juvenile) diabetes. Ben's and his parents' lives were changed forever. The boy eventually had to wear a glucose monitor. They had to check his blood sugar levels five times a day, and give him insulin injections before every meal. Wood had to set aside his career to care for their son full time.

A wealth of studies established that enteroviral infection may increase the risk of type 1 diabetes in children. An analysis of Japanese medical insurance records showed that after infection, a child was twice as likely to develop type 1 diabetes, a disease sharply on the rise in developed nations. A Finnish study suggested that the evidence provided "new opportunities for studying the viral etiology" of the disease. The gut provided a reservoir by which the virus could spread to the pancreas, hindering its ability to make insulin. In type 1 diabetes the immune system goes haywire, destroying the body's own beta cells, which means it can no longer process sugar. Another study likened the previous suspect viruses to the "judicial error" in Dostoyevksy's *Brothers Karamazov*.

It was not only an *Enterovirus*. Many type 1 diabetes sufferers linked the onset to bouts with much more common viruses such as flu or infections like strep, which seemed in some cases to turn the body not against the invader, but against its own pancreas, which makes insulin. What it meant, it could be said, was that our body microbes affected our health, and that viruses affected our body microbes.

The same could be said of human aging, which induces changes in the human gut microbiome. Our microbiome becomes less diverse and more proinflammatory as we age. On the hopeful side, the University of Washington's Matt Kaeberlein finished a promising study showing that the anti-inflammatory drug rapamycin could improve late-life health and the gut microbiomes of older mice, and some of the same benefits could seen in aging humans. The same was true of the diabetes drug metformin, and some biotechnology companies, including Unity, Elysium Health, Human Longevity, and the Google-founded Calico, were seeking to exploit these new understandings on the effects of the microbiome on the health of the elderly.

Can we somehow ameliorate the changes? Some studies suggested you could. An international consortium called ELDERMET was characterizing the microbes of the elderly. Older people's gut microbiomes had more of the undesirable Bacteroidetes and fewer of the desirable Firmicutes. The aging microbiome produced proinflammatory cytokine molecules, increasing our susceptibility to arterial sclerosis and neurodegeneration. Other microbe changes include increased dangers from colorectal cancers, colitis, and frailty.

Obesity, diabetes, cancer, autism, Crohn's, colitis, Parkinson's, multiple sclerosis, and aging were significant health issues, with built-in funding sources, in which microbes might play a role. Several researchers turned to looking at cystic fibrosis, and that brought them back to ancient environments.

Seeds of a lab and a food revolution

The MacArthur Fellow Dianne K. Newman, at Caltech, studied the coevolution of microbial metabolism with the chemistry of the Earth environment. "That's the core that links everything in my lab," she told me. She had several projects, the most important of which was the study of pathogens linked with cystic fibrosis — a direct outgrowth of trying to study how microbes interact with minerals. Microbes can use minerals to help catalyze their metabolism when oxygen is nonexistent. "The microbe doesn't care why the oxygen is nonexistent, to them it [the anoxic part of the human lung] is just a problem they have to solve," she said.

Studying bacteria in their environment had taken Newman from ancient sediments to modern hospital infections. "We have a really big problem in these infections because most conventional antibiotics don't work very well, at least [in] those patients with cystic fibrosis." Newman pointed out. "We have this vision for being able to iterate with analytical chemistry measurements of a habitat within the body, characterize it, so that we understood what constraints there are chemically." Once they had a better understanding of that framework, she hoped, they and others could go into the lab to design simplified environments to get at the microbial genes and understand what the bacteria need to survive. That approach comes straight from the world of the geoscientist: "We just import the tools the geoscientist uses all the time." In these chronic infections, the anaerobic body environments "almost looked like an ancient world in terms of their chemistry," Newman said.

Earlier she had spent a lot of time studying anaerobic respiration and anoxygenic photosynthesis, ancient processes that are still found on Earth today. From those studies she and her team became good at growing bacteria under anaerobic conditions and measuring their iron, which led to important perspectives on hospital infections and resistance to the medicines that treat them.

The biggest culprit in antibiotic resistance was not overprescription in human medicine, which was bad, but in animal husbandry. More antibiotics were given to animals than were prescribed to children. These antibiotics increased the prevalence of resistance genes in the environment. When the antibiotics remain in the food, they can select for resistance genes in the animals and in the people who eat them. These facts led Michael Pollan, a noted critic of the food industry, to write *The Omnivore's Dilemma: A Natural History of Four Meals* (2006), *In Defense of Food: An Eater's Manifesto* (2008), and *Cooked: A Natural History of Transformation* (2014), which were so effective that they in turn led to changes in the industry's practices. For nine years Perdue had been perfecting cheap and effective regimens for raising chickens without antibiotics of any kind. Approximately half of the chicken Perdue sells could be labeled "no antibiotics ever." Mortality rates and costs rose as Perdue removed antibiotics, but the company claimed it had made the effort over a fourteen-year span to keep their farms clean to reduce those factors.

Other companies such as Tyson and Foster Farms were also moving to factor out the use of antibiotics, but they continued to use ionophores, antibiotics used exclusively in animals in order to promote growth, prevent disease, and lower costs. The commitment by Tyson meant that, in theory, more than one-third of the chickens in the U.S. industry had been pledged to remove the use of "medically important antibiotics." Other major companies, including McDonald's, Chick-fil-A, Pilgrim's, Panera Bread, Chipotle, Whole Foods, and Applegate all announced that they would either reduce their use of antibiotics or had sworn off the drugs completely. But it was hard to assess the follow-through.

By 2017 it was understood that microbes could help to affect both our bodies and our behavior, train the immune system, affect our mental health, and even mold our intelligence. The Yale microbiologist Jo Handelsman, as White House Associate Director for Science, coordinated the attempts to apply the discoveries to medical treatment and policy. "Doctors," Handelsman told a press conference, "will learn how to tend ecosystems in the body as they do natural ecosystems in the wild."

In a three-day series of seminars sponsored by the Obama White House, the directors of the Center for Disease Control, the U.S. Department of Agriculture, the Food and Drug Administration, and even the Department of Defense pledged a five-part all-out collaborative "stewardship effort to track antibiotic use," including changes in food regulations, oversight by veterinarians of feed practices, a $20 million prize for a rapid diagnostic test to uncover new resistant microbes, and hundreds of millions of dollars to fund new research. A World Health Organization panel called for an International Anti-Microbial Resistance Registry.

A new term was coined, rebiosis, for nursing your microbiome, and the stories of people who changed their microbes to lose weight made it into presentations at conferences and home shopping networks. A Cornell professor of immunotoxicology, Rodney Dietert, claimed to have lost some thirty-five pounds by altering his microbiome. Microbes contributed to our cravings for dark chocolate, some argued, and they created metabolites that help process or produce key hormones, such as serotonin from some strains of *Enterococcus* and dopamine from some strains of *Bacillus*. Those 1,000 species in the gut, some 850 species on our skin, and about 300 in the mouth played a role in our weight, cravings, health, and even our potential longevity. The idea that the human microbiome played a role in chronic diseases such as asthma, autism, irritable bowel syndrome, allergies, depression, obesity, diabetes, colorectal and prostate cancers, psoriasis, and celiac disease reached a wider audience.

At New York University Blaser predicted that babies in the future would have at the first checkup a full-scale analysis of their gut microbes. The bacterial genes would be sequenced and counted and the urine analyzed for its microbial metabolites. A printout would dictate which microbes needed to be added to its diet, as either a probiotic or a prebiotic. Perhaps someday we would carry a chip or have an app to track our body microbes. The newest technology was producing results so fast that some 90 percent of the world's health-care data was created in the last two years. In the quest to commoditize the microbiome, technology was outstripping the ability to understand the biology. Someday, he suggested, we may even restore the feared *Helicobacter pylori* to its natural place.

It could be that such ancient microbes may rescue us from ourselves. Some of the claims, however, were spinning out of control. Then came a series of shocks—from Mars, and beyond.

12

Martian Chronicles

Killer Microbes from Outer Space

On September 28, 2015, the Washington, D.C., NASA conference room was packed. Under bright lights, the leaders of the Mars Reconnaissance Orbiter team squeezed behind a conference table as television cameras whirred and digital cameras flashed. The team announced it had found evidence of water on the red planet. Not eons ago, but today, dark, briny water was running in the mountain ridges around Hale Crater in subfreezing temperatures. In Nantes, France, the Nepalese student from the University of Arizona who first spotted the dark gullies answered questions via satellite. Perchlorate salts, like those found in earth's polar regions, or in rocket fuel, kept the water liquid. Pure water froze at Mars temperatures, roughly between 70 above and 100 degrees below zero Fahrenheit, but heavily salted water could stay liquid. "Water, albeit briny," announced John Grunsfeld, associate director of the mission, "is flowing today on the surface of Mars."

Calls poured into Nantes as the now–graduate student, Lujendra Ojha, headed from bar to bar to get a better cell signal. At each he had to drink a beer, "so by the time I talked to reporters from Argentina," he recalled, "and Bolivia, China, Hong Kong, India, and my native Nepal, my answers progressively worsened."

That scene came nine months after a similar gathering in San Francisco, when the NASA Mars Curiosity team announced that it had detected several emissions of methane, a signature gas created by microbes like Archaea, on the Gale Crater surface, lasting up to two months at a time. "It was an oh-my-gosh moment," the Curiosity researcher Chris Webster said. Then, late in 2016, the Mars Reconnaissance Orbiter detected an ice field the size of Lake Superior beneath the region called Utopia Planitia.

The announcements set off a search for microbes on the red planet and what they might mean to life in the universe. Liquid water, methane, life that could live off of sulfur and hydrogen, all seemingly telltale signs of potential microbial activity, helped to reignite interest in Mars in movies, nightly comedy, and social media. Millions followed NASA's twitter feeds. Mainstays like *Scientific American* and National Public Radio, buzz feeds, and web thought catalogs all commented. The number of hits on Curiosity's and NASA's Facebook pages soared, as did the number of its followers on Twitter. The NASA app called Be a Martian shared Curiosity's images within hours after they were received. Postdocs and graduate students tweeted at astrobiology conferences from Chicago to Nagasaki as real-time images poured in. I eagerly followed most of them.

The idea was that in its early years Mars offered a better prospect for life than Earth. When methane emissions were detected in two two-month bursts inside of Gale Crater, all the old arguments about Mars microbes returned. Some researchers still claimed that NASA's first lander, Viking, had detected Mars life in 1976. Early Mars had a wet-dry cycle that the early Earth lacked, and many thought that made it a likelier candidate for life's origin than Earth. The facts that the planet was smaller and that its lower gravity allowed rocks to burst into space when hit by comets, along with the fact that the closest large body was Earth, meant that pieces of Mars often landed here, bolstering the idea that our microbes may have originated as Martians.

As the Curiosity rover, a nuclear-powered Mini Cooper–sized vehicle, climbed Mount Sharp, the Russian Space Agency, with the European Space Agency, worked feverishly to finalize their ExoMars probe preparation for takeoff in 2016. The European Space Agency closely watched its Schiaparelli probe hurtling toward the fourth planet. A NASA probe catapulted toward Pluto while the expiring Cassini was purposely driven into a plume of water jetting off the surface of Jupiter's moon Europa. Around the solar system, a golden age of probes made the conditions of life look more possible in space.

How would you look for as microbial life on another planet? Hints came from Earth's most extreme and dangerous caves, its volcanic pools, and its coldest briny lakes. At the same time, new interpretations of microbe-human relationships took hold as new books appeared and commentators talked about the microbial cloud we carry around us. The business potential of microbial symbiosis brought us compost to clean wastewater and generate electricity. Microbes were projected to be a multibillion-dollar global business. From Jupiter's moon Europa to Saturn's volcanic satellite Enceladus, interest in the biosphere heated up.

Then Old Dominion University's Nora Noffke took a second look at Mars Curiosity's older photos and threw down a challenge. For three years, Noffke said, the NASA team had been staring at evidence of past microbial life on Mars, exactly the same structures she had found in Australia's desert or in the tidal marshes of Chesapeake Bay. The trouble was, they had missed it. It all began in 1976.

Panning for life

At midnight in a sweltering room in Pasadena in July 1976, Viking Mars team members sat hunched around a bulky monotone computer monitor, tensely awaiting the first data from the world's first successful Mars probe lander, the only Mars lander ever specifically designed to detect life. Over the next three hours each of Viking's first life-detection experiments came back with a striking signature. As the data fed in an agonizingly slow trickle into the Space Operations Facility, it was clear that carbon dioxide was released when organic compounds were added to Martian soil, though not when the mixture was superheated. This was a life signature, and exactly what had happened with the experiment on Earth. When water was added to the soil, oxygen was released, just as on Earth. The remote probe, panning for life, had found its signature in its first two experiments. The third experiment heated the soil, like warming food in the oven, and those results were mixed.

Arguments intensified, however, over the next several nights, as the fourth experiment's conflicting data came in. To claim life on Mars would be unprecedented. If they were wrong, no team member would live it down. Anything was better than striking out with a pompous grin on your face. Unbeknownst to most in the world watching, however, three of the four experiments on the primitive Viking lander could have been interpreted as testing positive for microbes, giving the same results as when they had been checked thousands of times on Earth. The researcher Patricia Straat told the director of the mission, Gil Levin, "That's life!"

However, the fourth experiment, employing a gas chromatograph of the kind employed by James Lovelock, and a mass spectrometer, a delicate instrument for measuring the size of molecules, showed no life—and not only that, but absolutely no organics—on Mars. That was a stunning result: organics exist all over space—on asteroids, comets, and meteors, and in interstellar dust. Not only that, but the experiment suggested that Mars's surface was poisonous or self-sterilizing. Mission scientists had a heated argument, and NASA eventually decided to err on the side of caution. The surface must be self-sterilizing, they concluded, owing to the planet soil's powerful oxidizing agents, which also helped to give it its red color. Viking had found a barren, windblown red planet, pockmarked by craters, cold and dead as the moon.

A few mission scientists disagreed, maintaining that the fourth experiment simply failed—as it had often in Earth tests. A group of activists, including Levin, wrote and spoke, goading NASA to release the full Viking data. On NASA's 2016 celebration of the mission's fiftieth anniversary, he repeated his call. He predicted Curiosity would find complex organics, which it did. When he saw the detection of methane bursts by Curiosity, he told me, he saw that the disappearance of the methane had happened too rapidly to have been caused by ultraviolet radiation: "This disappearance could have been caused by methanotrophs, which use methane and makes for a perfect little eco-cycle."

Other Mars probes gave conflicting results. NASA's Opportunity and Spirit rovers in the 1990s, whose daily reports thrilled millions of fans around the world, including me, were designed and constructed by geologists and engineers, not by biologists. Evidence of water came from the 2008 Phoenix lander, whose camera pictured clear droplets on its cold steel legs. Simulations suggested that either water condensed around windblown grains of calcium perchlorate, the salt-type mineral whose properties enable it to scavenge water from the atmosphere, or that the landing had stirred up dirty ice beneath the surface and drops had formed and melted on the legs. The point was, said the University of Michigan's mission scientist Nilton Renno, "On Earth, everywhere there's liquid water, there is microbial life." Such saline water on Earth housed microbes, in fact.

One of the best places to look for promising bits of Mars, paradoxically, was Earth.

Mars microbes on Earth

Walk along the frozen white expanse of the Antarctic and you will see bits of Mars, in the unmistakable form of small red stones. In fact, about ten pounds' worth of rock from Mars falls on Earth every year. If a large meteorite strikes Mars, it flings bits of rock into space beyond the small planet's gravity. As that planet's closest neighbor, our own, larger planet will find that its gravity traps some of the rocks, which fall to Earth's surface and are most easily found in barren, ice-covered regions such as the Antarctic. Their authenticity is determined by chemical analysis of their shock glass, the melted glassy substance original to the rock. If a stone's shock glass contains the exactly the same mix of gases as Mars's atmosphere, identified in the many Martian probes, it is from Mars.

Tiny squiggles on a famous Martian meteorite found in Antarctica, ALH 84001, in 1996 led the NASA researcher Dave McKay and his team to claim that they had discovered fossils of microbes. Today most regard that as unlikely. But eons ago abundant water had certainly flowed in Mars's oceans and rivers, the liquid's mineralized remnants clearly visible across the planet—floodplains, alluvial basins, even oxbow curves in long-dried great rivers. The original name given in 1887 by the astronomer Luigi Schiaparelli to the great rift valleys seen through early telescopes was *canali*, Italian for "channels" (though English-speaking researchers translated it incorrectly as "canals"). At the turn of the century, in Arizona, Percival Lowell thought he glimpsed active Mars rivers featuring seasonal changes in vegetation. In fact, numerous probes have imaged morning haze in Martian canyons. Seizing on Lowell's claim, Edgar Rice Burroughs, the author of the *Tarzan* books, wrote a wacky series of 1920s and '30s *Princess of Mars* science-fiction novels that sent generations of young Americans off to adventure. What Lowell saw was flaws in his mirrors. What Burroughs saw, as he divorced his wife for a Hollywood actress, was a gold mine of public gullibility.

Later NASA probes produced confounding results. The Mariner

probe in the 1960s strongly suggested that the Mars's thin, cold atmosphere permitted no possibility of pure liquid water, though the final orbiter clearly pictured the beds of ancient streams and oceans.

In 2010 a University of Arizona undergraduate studying the Mars orbiter images spotted intermittent dark streaks running in parallel at mountain ridgetops, then disappearing, as if in seasonal flow. What the Nepalese-born guitarist Lujendra Ojha noticed in the puzzling phenomena came from the high resolution imaging science experiment (HiRISE). The seeping dark streamlets appeared at dozens of sites. Excited, Ojha paired HiRISE observations with mineral maps of Mars. The spectrometer observations showed hydrated salts at several locations, but only when the dark features appeared and widened. Using the orbiter's compact reconnaissance imaging spectrometers, Ojha and the team then analyzed light reflecting off the streaks, detecting in it traces of either sodium perchlorate or magnesium perchlorate. Mars water contained a natural briny antifreeze.

Picture a cool planet with intermittent water and dryness and the largest volcano in the solar system. Giant lakes contain the same volume of water as that in the Arctic Ocean, fed by rushing rivers that deposit tons of alluvial sediment in their deltas. That was early Mars. Now picture a brimstone-smelling, acidic, ocean-covered planet with a toxic atmosphere and hot greenhouse gases, with neither oxygen nor radiation-sheltering ozone, that was pummeled by comets, then rammed by a Mars-sized planet, ejecting enough rock to form a gigantic moon that wrenched the planet's surface into skyscraper-high tides. Welcome to early Earth.

For that and other reasons, the NASA investigator Steven Benner and others suggested that life originated on Mars and was carried to Earth by ejecta. Poring over the old Viking transcripts in the Houston NASA library, Benner found precious clues in the forty-year-old transcript to what they had seen in the hot nights as the data unfolded. What he found was "mass confusion," he told me. Studying the DNA of ancient microbes and resurrecting their genes and proteins, Benner sought to connect the origin of life on Earth with the existence of life in the solar system. Mars, he pointed out in a series of papers that landed him in *Nature*, featured "warm temperatures and a wet dry cycle," he said, enabling amino acids to concentrate "for our kind of chemistry."

The trouble was that many past claims of Martian life or water had been so wrong. But many ancient microbes had thrived in similar icy, alkaline environments on Earth. For that reason researchers raced to sulfurous caves, hot springs in Kamchatka containing molybdenum and borate, Yellowstone National Park, and the brine lakes of the Antarctic. What they found was beguiling.

In the west

NASA's Chris McKay and Penelope Boston were two of the researchers searching remote, extreme outposts on Earth for signs of microbe metabolism and origins. A former New Mexico School Institute of Mining and Technology professor and the daughter of two circus trainers, Boston started off studying microbes in the Arctic and then switched to searching in deep caves. California's Alison Murray sought extreme microbes in the Antarctic. Suddenly, after Curiosity, everyone was interested in what could live in an ice-covered lake or plain, or in a remote cave or mine miles below the planet's surface. The likelihood of microbial life on Mars, which Boston had placed at 30 percent, was now, in her view, increasing. If tiny forms of life exist in hostile lake, cave, or mine environments, Boston reasoned, microbial life could eke out an existence on the subsurfaces of Mars.

Starting as director of the Cave and Karst Studies program at the Institute and cofounder of New Mexico's National Cave and Karst Research Institute, Boston made the case for Mars, helping to create the Mars Underground and putting together a series of discussions in which she won over the skeptics at NASA for life on Mars as a significant possibility. She was so successful that in 2016 the space agency named her the new director of its Astrobiology Institute in Moffett

Field, California, giving her the exciting opportunity to "help guide the science I love at a very high level," she told me.

From Nevada, the Desert Research Institute biochemist Alison Murray joined Louisiana State University geophysicist Peter Doran to study the microbes and ancient climates of the briny, ice-covered lakes of the Arctic and Antarctic, together finding a wide variety of Archaea and bacteria rising and falling with the seasons and times of day. "They don't do much—they hibernate most of the time," said Murray, who dipped ice cores down to the Antarctic Lake Vida's bottom, "but they're there." It was warmer further down below the brine, but the core tubes brought up ice from the deep sink.

That insight took researchers back to the American west, where the Berkeley biologist Jill Banfield studied the Colorado River and the water of an abandoned California mine at Iron Mountain. Banfield discovered dozens of new phyla of bacteria in a single toxic mine-waste site. The key was that these strange, previously unknown microbes depended on communities of other organisms to survive. This helped explain why so few could be grown in the lab. Working in the abandoned mine, Banfield's team applied a new technique to finding organisms and discovered several new phyla of bacteria, practically revolutionizing the tree of life. The Banfield team divided the 789 organisms into thirty-five phyla, twenty-eight of which were newly discovered, within the domain bacteria. They based the sorting on the organisms' evolutionary history and on similarities in their 16S rRNA genes; those that had at least 75 percent of their code in common went into the same phylum. The team found vastly different symbiotic species at each level and in each season.

In the fall of 2016 the Banfield team found new bacterial groups from a single Colorado aquifer, doubling the number of the planet's known bacterial groups, a massive underground find that again revised the tree of life. Banfield also studied the microbial colonization of gut of infants, taking her into neonatal intensive care units. Her work, and that of others in Yellowstone, in Chile's Atacama Desert, in Colorado's and California's abandoned mines, and even the inside of a dolphin's mouth, led to a new tree of life published in *Nature Microbiology*; about a thousand previously unknown species had been added over the past fifteen years. A second surprise in Banfield's discoveries was that almost half of the new bacterial diversity came from one group thought to live only in symbiosis.

Then Nora Noffke ignited the Mars search for life.

Hiding in plain sight

Through the hot coastal Virginia summer, Old Dominion University's Nora Noffke was studying the Curiosity rover photos from Gale Crater. The diminutive Noffke was a leading authority on microbially induced sedimentary structures (MISS), a term she coined for the stone patterns left by tidal microbial mats in briny tidal shallows. Many people were familiar with stromatolites, the sediment mounds deposited by ancient microbes, but few understood the significance of the flat tidal mats. Over a thirty-year career Noffke had traveled to five continents studying and categorizing a dozen typical mat shapes, from roll-ups to diamonds to waves and shelves. The moundlike stromatolites made tourist destinations in the shallows off Australia and Hawaii, and in the Caribbean. Noffke found her MISS in the extreme outback of Australia. Virtually no one else had seen them, and they so far ranked as one the oldest pieces of evidence of life on our planet.

NASA invited Noffke to speak at a meeting to choose the landing site for its 2020 Explorer. If early Earth and Mars were similar, Noffke said to the group, perhaps there were microbially induced sediments preserved on the planet. In the audience, the Caltech geochemist Ken Farley sat up in his chair. His team had just published a paper on the polygonal structures seen by Curiosity in the ancient mud of Mars's Sheepshead Formation, a coastal plain on the route as it traveled twenty-two kilometers to Mount Sharp. After the meeting Farley sent Noffke the image and asked her what she thought. She studied the photos taken on Sol 126, the Mars day that Curiosity was in Sheepshead. Her heart leapt.

The images looked familiar. She spent hours making a drawing and sent it to Farley, saying, "This could be microbial. Should I analyze them further?" He said yes.

Noffke started her study in May, then, after teaching ended, resumed in July and August. All through the summer piles of papers stacked up higher and higher in her home office. She knew the dangerous tendency to see microbial structures everywhere. She studied her own personal huge archive of MISS photos from Earth. In the afternoon, when the sun lowered, she walked along the tidal flats of Chesapeake Bay, thinking about the ancient ocean beds on Mars, with its tinier sun, similar day length, and warmer temperatures. She was a Mars beginner. "I'll submit a hypothesis paper and see what people say," she thought.

When, in January 2015, she published "Signs of Microbial Life on Mars," the Curiosity team reacted strongly. Noffke was seeing the equivalent of "clouds in the sky," said one scientist. The team created a website to refute Noffke's claims.

But when Curiosity's research found perchlorate salts in briny water below the Martian surface, Michael Meyer, the lead scientist for NASA's Mars Exploration Program at the agency's headquarters in Washington, gave a boost to the search for life: "The more we study Mars, the more we learn how life could be supported and where there are resources to support life in the future." Then came the news of the methane burst. Yes, the methane bursts could have been contamination from the Curiosity. Then came the briny water.

The rancor grew. The Curiosity team said she had the geology wrong. Noffke responded. "It's now an eroded hillside but it has been a former lake, in a completely different paleoenvironment. They say it's a braided river. That's simply not true. It's a hillside left from a meandering river system. That's exactly where you would have microbial mats on Earth, in places like that!"

The only way to settle the question was to put people on Mars. The big problem was carrying enough fuel for liftoff from the Martian surface so that they could return to Earth. For that reason, a first step would likely be humans in orbit around Mars, as proposed by the U.S. Planetary Society, presided over by Bill Nye. Looking further, NASA was planning to send humans to Mars as early as the 2030s, while the European and Russian ExoMars mission, scheduled to land in 2018, was to select a lake bed for its site.

Further evidence of ancient microbial activity in the Gale Crater basin came from the University of Oregon, where the geologist Greg Retallack noted the soil's high level of sulfate, which could only be produced in an anoxic environment by anaerobic bacteria. Some of the "vesicular structures," or bubbles, in the Curiosity photos resembled those produced by microbes on earth after a rain, Retallack wrote in *Geology*. For her part, Noffke felt bruised by her experience with big-ticket science and the public fascination with extraterrestrial life. Her paper was merely a hypothesis, not a full-blown argument or claim, and the hostility of the Curiosity response took her aback. The Curiosity team did, however, shape a new itinerary for returning to the site of the methane burst, at exactly the same season as the first was detected. Perhaps it showed, more than anything else, how important the research was becoming to a popular base that included Twitter and Instagram followers like me.

For that reason, researchers and fans eagerly followed the European Space Agency probe Schiaparelli as it winged into orbit around Mars in the fall of 2016 and prepared for its test rover's descent. Monitored by its orbiting parent unit, the Schiaparelli test lander descended into the Meridiani Planum region on October 19, 2016. The parachute deployed twelve kilometers up, and the heat shield was released at 7.8 kilometers, as planned. Then a mistake happened with its inertial measurement, which ran one second too long as it became saturated with data and generated an estimated altitude that was below ground level. The one-second error triggered the second parachute to fire too soon, as did its braking thrusters, causing the dummy lander to crash hard and disintegrate. The debris could be seen from its orbiter and from NASA's Mars Explorer Rover.

The test was a bitter disappointment, but Schiapparelli's real rover landing was not scheduled until 2018. By then many will be looking beyond Mars, with its intense solar radiation and sterilizing soil. How would you recognize life on a solar system body if you saw it? What exactly would you look for, and how? There was no reason to suppose that either Earth or Mars was the only, or even the most ideal, model for life. Many thought that good places to look might include the moons of Jupiter. And Saturn.

Red rain

"But what exceeds all wonders, I have discovered four new planets and observed their proper and particular motions, different among themselves and from the motions of all the other stars; and these new planets move about another very large star, Jupiter," Galileo wrote to the Tuscan court in 1579, shortly before his treatise *Sidereus Nuncius* (Starry Messenger) gave the world his full report on a sun-centered system of the heavens. Galileo accurately mapped Jupiter's four moons and charted their orbital periods, size, and day lengths. Every amateur grade-school astronomer ever since, equipped with a good telescope and thirty nights of viewing, could do the same. The revolution that followed put humanity in a more accurate and humble relation to the universe.

To modern microbe hunters, those moons of Jupiter and Saturn held special allure. Two—Saturn's Enceladus and Jupiter's Europa—stood out from the rest, for different reasons. Enceladus had volcanoes, organics, water, ice, and showers of volatile hydrocarbons pouring down as red rain. Europa offered hints of a vast ocean.

It was not just any ocean, it was potentially liquid, heated by the squeezing of the moon by Jupiter's massive gravity, like a tennis ball, so much so that it cracked the thin ice of the surface. A tumultuous NASA mission, aptly named Galileo, launched in 1989. The vessel spun into the giant planet's orbit from 1996 to 2003 and soon began sending back amazing images of planet and its moons. What struck team scientists were the fissures in its waterlogged moon, Europa, showing that its ice was not so impermeable as they had thought. Rather, it was thin, and water may seethe underneath, presumably heated by a molten core, even though the surface temperatures range around 130 degrees below zero Fahrenheit.

Most likely Europa and Enceladus had volcanic magma heating their liquid water. They probably featured exactly the kind of hydrothermal vents that so excited Earth researchers as a source for Earth life. Then it was seen that Europa featured sulfate salts, much like the lakes in Antarctica. The significance was that life on either of those places would be proof of a second genesis, independent of the Mars-Earth axis.

Even more stunning, the Galileo probe seemed to show water gushing from Europa's surface into outer space. When arguments erupted on the Galileo team, a younger member decided to turn down the exposure on the black-and-white images. Suddenly water plumes could be distinctly seen shooting out from the ice, hundreds of miles into space. Subsequent Hubble images seemed to confirm the plumes. Europa's icy surface was riddled with undulating cracks and wriggling fissures. Beneath its frozen exterior distant, icy Europa appeared to feature a liquid ocean.

Then came Saturn's Enceladus, which had the same geysers as Europa, but also a torrential pouring of particulates that made the rock-and-roll musician Peter Gabriel's song "Red Rain" almost visible. Enceladus was so enticing as a target for life that, in the last months of 2015, NASA's Saturn Cassini team planned a daring maneuver with its dying probe. They drove Cassini through an Enceladus plume as it was ejecting into space and analyzed the content—the first time an alien ocean had ever been sampled. What they found was very similar to samples taken from Earth oceans: present were hydrogen, carbon dioxide, methane, and tiny silicate grains possibly produced at alkaline hydrothermal vents.

Add in Saturn's moon Titan, where Cassini uncovered chemical sig-
natures suggesting conditions suitable to a primitive, exotic precursor to methane-based life. Hydrogen molecules flowed down through
Titan's atmosphere and disappeared at the surface. A paper in the
Journal of Geophysical Research mapped hydrocarbons on Titan's surface and found no acetylene. This absence was important because that
chemical would likely be the best energy source for methane-based
life, said Chris McKay, at NASA's Ames Research Center. One interpretation of the acetylene data is that the hydrocarbon is being consumed as food. But McKay said the flow of hydrogen was even more
critical. "Hydrogen's the obvious gas for life to consume on Titan,
similar to the way we consume oxygen on Earth," McKay said.

After the European Space Agency Rosetta orbiter landed on the
comet Churyumov-Gerasimenko, and after a nine-year-old NASA
probe to Pluto showed it to be covered with water ice and erupting
mountains, the solar system became a more interesting place. The
Japanese space agency, JAXA, approached NASA about a joint mission to bring back a sample of water from Enceladus. Amazingly, the
NASA probe discovered that an ocean of liquid water might lie beneath Pluto's frigid ice surface. Then came the discovery of seven
Earthlike planets orbiting a nearby red dwarf star, which are three
times more numerous in the universe than sunlike stars, and the
search for signs of extraterrestrial life was reinvigorated.

Taken together, the islands of our own solar system were soggier,
potentially warmer, and richer in minerals than expected. But what
kind of life might they have? To answer that, a whole new system of
interpretation was called for. It may come from one of the iciest lakes
of the Antarctic.

Second genesis

It was unseasonably warm in the Lake Vida valley when they built
their camp. The laconic Alison Murray was there with Peter Doran,
who had made the same trip to the Antarctic many times before. They

had eleven hours to get their camp up before the helicopter had to leave. The wind howled and the temperature plummeted in the harsh light of the vast, dry valley.

They planned to drive cores to the depths of Lake Vida, in the southernmost McMurdo Dry Valley, a frozen lake isolated under year-round ice cover, with water that was saltier than seawater. They had brought Lake Vida to the public's attention a few years earlier when they discovered that a primitive form of cyanobacteria, frozen in its ice cover for more than 2,800 years, could be thawed and reanimated. Anything else that lived down there had been sealed off by ice for millennia. A nearby plain they found, by contrast, to be completely devoid of life — the only such region on Earth. The wind whipped up, and Murray tightened her gloves.

Raised in California, she had surprised herself out here in the frozen wasteland. First, she found it beautiful. Second, she loved the intense cold, light all day in summer, only shadowed for a couple hours at 3 A.M. by the mountains, and dark all day in winter. Here, "with life signals so weak we could barely detect them," she told me, she could truly understand microbes in an integrated system with chemistry, geology, and climatology, a strange icy world she came to know more intimately than her own. She had come back now to dig deeper.

They thought the briny lake might give clues to the types of microbes that could exist on Europa or Enceladus, or even Titan. Then they saw that the water contained perchlorate salts, made famous on Mars, in the heaviest concentrations seen on Earth. They drilled down deeper and deeper, to a depth of twenty-seven meters, and later radar analysis showed the brine could go as deep as fifty meters. It was strange that the surface was frozen solid but the depths were warm enough to support liquid water.

The team pitched their small, bright-orange sleeping tents in neat rows, setting up a clean room on top of the hole, wearing the type of white suit used in germ labs to keep conditions sterile. When one microbe they discovered became of medical interest, Murray was contacted by an immunologist seeking promising compounds in its metabolism. Now they set to work. In this frozen time capsule for ancient DNA, they wanted no mistakes.

The first Antarctic season proved to be unseasonably warm, and meltwater was flooding down by the time they were half-finished at the lake; they raced back to the tents at full speed to move them to higher ground. The wind sheared their faces, and the core samples tugged down the sleds. They needed to finish as fast as possible.

For three weeks they sampled, storing the samples to go for further analysis to McMurdo and labs in North America and Australia. They were in contact with McKay in California. Murray had studied at the University of California and worked with one of the pioneering researchers of Archaea. McKay had grown up in Florida, learning physics by tearing around the dirt back roads on his motorcycle. The Viking images had changed his life. He thought hard about a second genesis of microbes on Enceladus. Life might not have RNA and protein and DNA. McKay also asked if life had to be water based. Oceans of hydrocarbons could be much more efficient energy producers than were water oceans. It was McKay who had advised that they bring a better drill than the one they had planned to use.

The microbe inhabitants of Lake Vida, Doran and Murray found, lived in total darkness in temperatures well below freezing, yet the lake teemed with unbelievable diversity, the more so the deeper you went. "Actually, the water temperatures rise the deeper you go," Murray told me, of the depths that shade a little closer to Earth's molten core. "The conditions are much like we think could be true of Enceladus or Europa." Then the warm temperatures created a dense fog that settled in. With only Murray and one team member left, the rescue helicopter they were desperately awaiting could not take off from McMurdo.

They were gradually running out of food. They sat and waited as the wind howled and traded stories until they were bored and bargained for chocolates.

At the end of his career Stanley Miller, escorted to conferences in a wheelchair, became convinced that life's building blocks might have been most effectively concentrated in ice. Early Earth might well have been quite cold in regions for long periods of the year, since the early sun generated some 30 percent less energy than it does today. Miller and his students took the same vials of chemicals he had worked with during his entire career. They had been frozen for years, yet, when thawed, the building blocks of nucleotides formed much more rapidly than at room temperature, owing to the molecular properties of water ice.

More intriguing, said McKay, the key to icy Enceladus, Titan, and Europa was that if life was there, it had to start there as a separate genesis from those of Mars and Earth. Some similar evidence had piled up from the forbidding Barberton Mountains in South Africa. It seemed as if life could indeed have started in ice. Deep inside the icy water, Murray extracted samples of genetic material and found a thriving world of life, moving very slowly, almost imperceptibly, but fashioning its miracle of RNA, proteins, and lipids from the energy sources of hydrogen, sulfur, nitrogen, iron, methane, and carbon.

Finally the helicopter landed and they raced to load. After numerous trips to the Antarctic, Murray was hired on to the new NASA team contemplating a Europa landing.

The potential revolution in these cold-world studies was that Earth's microbial life was much more diverse and opportunistic than we had imagined, and that life elsewhere could be more so. It seemed as though life could do anything. If that was true, then microbes could be engineered to make new energy, clean up the plastics choking our ocean, or perhaps deliver our medicine in such a way that pathogens could not resist. A vast effort in understanding and even making synthetic microbes was gaining steam. One place it took off was in the mountains north of Silicon Valley.

Something to talk about

Towering above San Francisco Bay, the Lawrence Berkeley National Laboratory was a hotbed of metabolic engineering as interest grew in symbiosis as a new paradigm for life. Metabolic engineering, predicted by the lab's director, Jay Keasling, could provide alternative medicines and fuels. Berkeley's John Coates studied microbial metabolism of industrial and military perchlorate to pioneer better waste treatment. Then it turned out some of that same perchlorate filled the soils of Mars.

Some of the research, said the University of California, Santa Barbara's Irene Chen and others, could increase interest in the work on phage therapy, a different model for controlling infection by using the age-old viral scourges of bacteria. "The problem of antibiotic resistance is becoming personal to me," said Chen—she was expecting her first child. In 2016 she won a $2.1 million National Institutes of Health New Innovator Award to sequence the phage viruses from a nearby medical facility. At her California campus she was excited to see that Jeffrey Bada, once a postdoc with Stanley Miller, had kept the famous, original origin-of-life Miller-Urey apparatus intact. "We marveled that it was still on," she recalled. "Jeff was boiling coffee in it!"

The Berkeley researchers led by Keasling discovered they could engineer the common yeast *Saccharomyces cerevisiae* to make high-energy, fatty acid–derived fuels they envisaged could be scaled up for industrial production. In American agriculture, the giants Bayer, DuPont, and Monsanto were investing a combined total of $1 billion in natural microbial pesticides called biologicals to increase yield and control pests.

One of the initial successes included a bacterium, discovered by researchers at the Kyoto Institute of Technology and Keio University, that could completely degrade plastics in the ocean. Previously only fungi were thought to be able to do so, but Shosuke Yoshida and his colleagues found a bacterium called *Ideonella sakaensis* that produced two plastic-degrading enzymes. The process was slow—it took up to six weeks to degrade a small film—but the team hoped to mass-produce the enzymes directly. Another team at University of California, Santa Barbara discovered that human anaerobic gut fungi that could degrade plant waste.

In 2016 the Venter Institute reported making the simplest living cell ever seen, with only 473 genes, nicknamed Synthia 3; still, they were unclear what one-third of its genes did. "We don't know a third of the basic knowledge of life," Venter told the *Atlantic*. They raced to uncover those functions, but in the meantime, the same technology "could be used to construct new cells with desired properties," commented the University of Valencia's Rosario Gil, "an unprecedented step."

One of the greatest of microbial inventions was photosynthesis, extracting fuel from sunlight and water—a feat very difficult to accomplish by any artificial means researchers raced to devise. By November 2016, however, a University of Illinois and Berkeley team led by Krishna Niyogi and Stephen Long had engineered 20 percent more efficient photosynthesis in tobacco plants by reducing a cooling mechanism of chloroplasts that protects them from harsh sun. "Making plants that yield more," a Munich-based molecular biologist told *Science*. "That is something that everyone should be happy about."

Business interest in symbiosis as a new paradigm in life science became more widespread. The number of companies selling targeted probiotics or synthetic microbes, such as Emefcy and Syngensis, was soaring. The number of researchers studying microbes in waste remediation was growing, as was investment. Microbes, it was seen, can survive in the stratosphere and in space. They might have survived in the pockets of comets, as Canadian researchers found. They might lie at the sites of asteroid impacts on Mars.

A prospect for better space probes was raised by the eminent physicist Freeman Dyson, commenting on the increase in privately funded cheap spaceships and space tourism. Future planetary probes could use the best new information technology at a fraction of the cost of NASA probes, Dyson wrote, and would be launched in quick succession, so that one failure need not set back the search for life for such a long time.

By 2016 a host of researchers had penned a global call for an international human biome effort, including Nicole Dubilier and Margaret McFall-Ngai. Noting that half of the world's oxygen is produced by microbes at sea, that we had a huge collection of their genes but only knew what half of them coded for, and that microbes did not observe national borders, they proposed combining the efforts in Japan, Europe, the United States, and Canada to create a global effort. "The revolution in understanding," they wrote, "must be seized."

As attention increased, researchers turned to the recesses of the human body, to insect bodies, and to remote mountaintops and saline lakes, seeking more clues. What no one expected was that the discoveries on Mars and in the solar system would give new life to longtime research on life in similar environments on Earth. As the interest increased, so too did the controversies.

"I have written the wonders I have seen," concluded Galileo, shortly before he was placed under house arrest, "and of these discoveries more news will follow."

13

Sustainability

Toward a Microbe Economy

The hot sun blazed over California's arsenic-ridden Mono Lake as Duquesne University's John Stolz motored to shore with sample buckets of its salty water. Strange towers, caused by the mixing of hot springs and lake water, loomed like ghostly hands in the blazing light. Salt coated his sweating face. No fish swam in the lake, so ethereal that it had been made into a Pink Floyd album cover. Tendrils of brine shrimp and brine flies fed on the microorganisms in the extremely alkaline water, comparable to the produced water from the shale gas wells of Pennsylvania, where the sixty-one-year-old Stolz advised state and local governments on the microbial remediation of toxic waste. "Fracking waste is almost exactly like Mono Lake water," Stolz told me. "Toxic. But microbes love it."

Microbe behavior and feeding underlay the growing industries of bioremediation, bioreactors, energy production, and bio-recovery, helping to clean oil spills, spent nuclear fuel, heavy metals, toxic chemicals, and mercury and pesticide runoff. Using the tools of high-speed data analysis, which can track gene expression in multiple organisms at once, researchers studied the molecular pathways by which microbes liberate energy from almost any chemical bond. Microalgae and fungi could digest heavy metals. Some sulfur-reducing bacteria could neutralize radioactive waste. Extreme bacteria and Archaea, marketed by companies such as Heartland Microbes, were used in secret industrial recipes that removed pollutants like chlorine, petroleum, and bromine. A Vanderbilt University team discovered an archaeal enzyme that could kill some of the most resistant pathogens, including MRSA, anthrax, and C. diff.

In the Gulf of Mexico research continued into the communities of natural oil-eating microbes such as Thalassaleituus olieverans, Neptunibacter, and Alkanivorax borkumensis, which seemed to clean up much of the Deepwater spill, albeit slowly and with a great deal of disagreement about effectiveness. In South Africa the international team of researchers working two miles below the Earth at the Mponeng gold mine had uncovered a "veritable zoo," said Ghent University's Gaetan Bogonie, of microbes that remediated radioactive heavy metals. He predicted an explosion of interest in commercializing the discovery, with an upcoming paper about to detonate ideas on where that extreme life came from—much of it being found in deep seawater, perhaps where nascent life had migrated to survive the primeval heavy bombardment of meteorites. "That means we are all immigrants," the bearded Dutch scholar said to me, referring to the rising political intolerance in many Western countries.

With some 72,649 microbial genomes sequenced, the U.S. Department of Energy expanded its support of microbial genome sequencing for other industrial and energy uses. Biofilms of marine and sewage sediments could be used to induce an electrical current, reported the U.S. Naval Research chemist Leonard Tender. With new tools that could help us understand the interaction of geology, genes, and behavior, we were on the verge of a revolution, taking advantage of a new understanding of the power of microbes.

Investment in synthetic biology, the engineering of microorganisms to generate fuel, diagnose disease, or remediate waste, was now expected to rise as much as 230 percent and reach $38.7 billion a year by 2020. Small companies such as Adaptive Symbiotic Technologies sold fungi to help plants beat heat, cold, and drought. Large companies such as Monsanto claimed to be "rewriting agricultural history" by creating crops to increase yield or prevent disease, while also fighting for public acceptance and government approval. The market included synthetic DNA, medicine, pesticides, and food, as well as the technologies to create or modify them. Key players included Agrivida, Evolva, DuPont, Exxon Mobil, and Biosearch Technologies, as well as biofuel companies like Coskata, Amyris, and Mascoma; Paris, Ghent, Boston, San Francisco, Tokyo, Beijing, and Singapore were some of the key sites of innovation.

Microbes were present virtually everywhere on Earth and survived in conditions of extreme acidity, temperature, pressure, and toxicity. Microbial communities thrived deep down in fracking wells. Could a microbe revolution deliver on a promise to clean up toxic waste and promote a renewable food supply? Perhaps the reason we had not taken fuller advantage of them before was that we had never looked closely enough.

Arsenic and white lace

In 2016 John Stolz, director of the Duquesne Center for Environmental Research, realized that "all these cool things that Lynn Margulis put in my hand are converging," as he told me, excited, from his cramped Pittsburgh office. Mono Lake water resembled gas drilling waste and the brine on Mars, and his work on arsenic-metabolizing microbes could help us confront a food crisis. It all began in 2010, when Pittsburgh drinking water was in trouble. One potential culprit was fracking. "Companies were allowed to take produced water from wells to public treatment plants, but nobody did the math because the stuff was way salty," Stolz said. In 2012 a state moratorium was instituted and new regulations put in place, "and we thought it was a done deal." But it was not.

Part of the problem was that the deep Earth microbiome was still such a mystery. "You really need to know what's going on," said Stolz of the contaminants in Pennsylvania. "We didn't know where the bacteria were coming from, other than from the water produced from the wells, but it could have been Texas or Oklahoma, where the rigs were waking up things in subsurface, one to one and a half miles down." There was life down there. "They're missing space and water, and fracking opened up space and water," Stolz observed. "There is a whole dark ecosystem miles down in those shales!"

The former aide to Margulis suddenly found himself giving inter-views and being deposed by lawyers as he analyzed the microbes of the Marcellus Shale for their ability to clean up the waste of oil drilling. It turned out that Mono Lake provided a glimpse of which microbes might live on Mars, metabolizing perchlorate. "I love it," he said of the strangely beautiful stop for millions of migrating birds. "Nothing there but microbes, brine shrimp, and brine flies. It's a unique place." When his colleague Ron Oremland sent him a Mono Lake sample, his lab manager "thought it was contaminated water from the Monon-gahela River! The sample just said 'Mono' on it." The only odd thing about Mono Lake was the lack of acidity. "The pH is 9.8, and the pH of fracking waters is lower," Stolz said, "but we're seeing some very simi-lar microbes." Those halomonads of Mono Lake are similar to those in drilling wastewater.

Suddenly the strange, obscure microbes in Earth's briny waters became the targets of intense interest. For twenty years Stolz and U.S. Geological Survey senior scientist Ron Oremland had studied bac-teria that stripped chloride off organic molecules, that could live by breathing and metabolizing arsenic in brine. "The key to Mono Lake was the fact that an Arizona group confirmed that microbes there did in fact use arsenate and chlorate to grow," he said. Suddenly there was a "chance of microbe fossils" in Martian rocks, said Old Dominion's Nora Noffke, that lived off the same dense, nasty brine. Stolz laughed at the turnabout. "It meant we hadn't been toiling all these years in vain!"

His work led the city of Pittsburgh to explore bringing bioreactors to generate its electricity, burning methane produced by human, food, and animal waste. Other nearby states followed suit. One Cleveland company made natural gas from the food waste at the Indians' Jacobs (now Progressive) Field. Two facilities in Wooster, Ohio, generated electricity from wastewater to run the waste plant and, partially, the drinking-water facility. Epcot Crenshaw, a company based in Pennsylvania, also assisted in the study by Duquesne. "The fascinating thing is, an MBA student crunched the numbers, and based on cow and pig waste we could generate eight trillion cubic feet of natural gas a year, which is a third of our current production. Renewable, sustainable, clean," said Stolz. "There isn't a municipal wastewater-treatment plant in this country that shouldn't be generating their own wastewater energy!"

Arsenic became a danger in baby food as we learned that food plants like rice concentrated natural arsenic in the soil, potentially leading to dangerous levels in rice-based baby food (arsenic, like lead, can damage a growing brain). Suddenly it turned out that the Stolz-Oremland study of arsenic-based metabolites had health applications, leading to a ban on arsenic in poultry feed. Microbes concentrated feed arsenic in waste to move easily into drinking-water resources. Microbial arsenic from natural or manmade sources is potentially affecting "tens of millions of people worldwide," Stolz said. The race was on to turn microbes into remediators of our folly.

Sympathy for the devil

A series of institutes and companies latched onto the promise of microbes as possible remediators of toxic waste. The Earth was awash in a flood of plastic, used fuel oil and gas (FOG), wastewater, and human and animal waste. A host of companies vied to turn that free matter into energy. The Israeli company Emefcy moved its headquarters to Australia and signed deals in China and Africa to generate energy from wastewater. In Oregon, EcoVolt was marketing microbial fuel cells to developing nations in Africa.

Although it may seem insignificant, something as common as waste had the ability to dictate future energy advances. Most people viewed it as figuratively and literally beneath us. But the organisms in waste could take its most toxic substances and turn them into energy. That ability was not new. They had been doing it for nearly four billion years.

The Earth four billion years ago was a toxic planet. Volcanoes spewed poisonous gases into a noxious atmosphere. The environment in which bacteria flourished was a chemical cesspool of lethal carbon monoxide, hydrogen sulfide, and hydrogen cyanide, containing little or no oxygen. Temperatures under a black, ozoneless sky ranged from the broiling heat of lava flows to the frozen ice fields at the poles. Radiation levels were high.

Most early microbes could not use oxygen, though once it took hold a few of them developed ways to live in a breathable world. These two-faced "facultative anaerobes," observed J. C. Hayes, of Heartland Microbes in Michigan, could fix nitrogen and break down numerous poisonous compounds in soil and water. They could also remediate many forms of our most difficult industrial waste. Heartland Microbes, and companies like it, marketed microbes remarkably "similar to the first primordial mixture . . . found on Earth," they claimed, including anaerobes (those that could not tolerate oxygen), aerobes (those that require oxygen), and facultative anaerobes (those that can survive in either environment). In a liquid culture, the company's "special blend" could reduce hardpan-baked dirt salts and remove fertilizers, diesel fuel, gas, oil, pesticides, and hydrocarbon spills.

Many companies selling microbes to remediate dirt came from the U.S. Rust Belt. Microbe Inotech in St. Louis used microbes containing electron-donor chemicals to turn toxic waste into energy. Service-Tech in Cleveland used microbes to clean industrial heating and air-conditioning ducts, commercial ovens, industrial exhaust systems, and municipal groundwater. For its part, Heartland Microbes featured microbes grown under a state of starvation—as they would have lived on the early Earth. The organisms that would immediately, on application, devour petroleum included cyanobacteria and microbes that can eat benzene and *Bacillus*, *Pseudomonas*, *Azotobacter*, *Azospirillum*, and *Rhizobium* species. Shell Oil and British Petroleum also used them. Oil-cleaning microbes were a $50 billion business in the United States.

A series of conferences encouraged students to come up with new ideas for microbial remediation. At the Genetically Engineered Machine Competition (GEMC) in Delhi, a team of Chinese students unveiled a "three-species" microbial fuel cell that could match a lithium battery for eighty hours, at little cost and with no pollution, producing a stream of 520 millivolts, using *E. coli*, *Shewanella*, and *B. subtilis*, all on a spoonful of sugar. The ten-year anniversary of the Boston-based iGEM student conference competition featured teams from Bandun, India; Manaus, Brazil; Hong Kong; and the Netherlands (the last dressed in banana costumes), each offering ways to reengineer microbes to do our bidding.

It was a new frontier, filled with big and sometimes spurious claims, and well intentioned if occasionally disappointing findings, where some promising avenues stalled and the results of some experiments could not be reproduced. Northeastern University's Slava Epstein returned from Greenland crestfallen in the fall of 2016, because "if you take two centimeters of soil, even one next to the other, they had totally different microbial ecologies," he told me. The Boston company SERES's microbiome-transplant pill failed in phase II clinical trials, causing its stock to plummet. Some researchers confused correlation and cause and effect. Some microbes were beneficial, some were harmful, and some were both harmful and beneficial, in differing amounts under different conditions.

Other companies marketed microbes to clean up oil spills in garages, in boat docks, and in your backyard. Suddenly, some age-old remedies became big business. Then all eyes turned to medicine, and a known but often-ignored natural defense against pathogens.

Phage-turners

As the crisis in antibiotic resistance continued to build, a few synthetic-biology researchers turned to bacterial viruses as possible replacements for antibiotics. Their research focused on phages — strange, primitive, creative viruses that infect bacteria. It had long been known that phages, on which lab research was active in the 1940s and '50s, could treat infections by attacking pathogens. The German and Russian armies included phages in their medical kits during World War II. The early research into DNA involved an active effort in phage study. Suddenly, in 2016, the phages made for hotbeds of interest in labs like that of Irene Chen at the University of California, Santa Barbara, who helped to pioneer much of the work of her mentor, Jack Szostak, on synthetic biology.

An English researcher first observed phage activity in 1915, but the French-Canadian microbiologist Félix d'Herelle gave the bacterial viruses their name in 1917 and tested their therapeutic potential on chickens and cows. Eventually human tests were conducted, and Eli Lilly commercialized phage therapy in America during the 1940s. Phage kits were used to treat German and Soviet soldiers suffering from gangrene and dysentery. The development of antibiotics in the 1950s, however, set phage therapy back, as did the gene revolution; but suddenly, with antibiotics in retreat worldwide, the phages made a roaring comeback. On the hundredth anniversary of the discovery of phages, researchers gathered in San Diego for the 2015 "Year of the Phage."

It looked as though it would turn out to be a promising avenue of research. The success rate of phages when administered topically, intravenously, or orally ranged from 80 to 95 percent, with few side effects, on such pathogens as *E. coli*, *Acinetobacter*, *Pseudomonas*, and *Staphylococcus aureus*. Phages also could be used against bacteria containing NDM-1, a gene that is one key to antibiotic resistance. Skin grafting for wounds, burns, and skin cancer could also be improved by using phage therapy to prevent infection. Some experiments uncovered phage antitumor agents. Finally, because bacteria cause food to spoil, phages were studied for their potential to prolong freshness. Amazingly, wastewater from sewage systems was one of the richest sources of promising phage strains.

Ten million trillion trillion or more phages inhabit the Earth, more than all free-living organisms combined. Vancouver researchers claimed phages cause a combined trillion trillion infections *per second*, destroying 40 percent of all ocean bacteria each day. As the carbon-containing bacteria die, they sink, removing one large portion of greenhouse gas from the air.

But the tiny viruses, almost on an atomic scale, are very hard to grow, and the techniques for analyzing them are in their infancy. Still, argued the University of California microbiologist Forest Rowher, editor of the book *Life in Our Phage World*, they offered the potential "to hack the human microbiome," as he told the *New Yorker* blogger Nicola Twilley. "You are going to see people manipulating individual bacteria species in someone's gut before too long, because it's not very hard to do."

The bottom line, Rowher told me, as he rode in a boat studying coral reefs off the Caribbean island of Curacao, was that phages will probably not be used on their own but rather to weaken resistant pathogens before attacking them with conventional antibiotics. "It's an age of discovery," he said, echoing other researchers, as the boat motored along. "The cell is not the fundamental unit of biology. It's [the] phage!"

At the same time, at Harvard, the Wyss Institute synthetic-biology team, headed by Pamela Silver, was working on detoxifying a common pest. Her team reengineered one of the most annoying, and harmful, of gut microbes, *E. coli*, to be able to tour the gut and monitor the environment for pathogens by inducing changes in the gene expression of intestinal microbes. Interest deepened so much that she was invited by the insurance company MetLife to lecture in a special panel on extending healthful human life spans.

The international microbe consortium was going strong. Forty-eight scientists from fifty institutions in the United States formed the Unified Microbiome Initiative Consortium (UMIC). Then the consortium expanded with the National Microbiome Initiative, announced in 2016. The scientists envisaged a coordinated effort spanning national, institutional, and governmental agencies to support the goal of driving cutting-edge microbiome research, enabling breakthrough advances in medicine, ecosystem management, and sustainable energy. At the same time, discussions of a global medical microbiome initiative intensified in government policy offices around the world.

The business and government interest in the microbiome was growing. Then some researchers turned to global warming, and the microbes suffered a reverse.

Global warming

The snow is falling on Easter in Avon, Colorado, the wind whipping across the top of the Birds of Prey giant slalom, where I pivot madly to miss the looming, frosted pines. Every time I come here, I think of the microbes on which snowflakes may form. Microbes float in the stratosphere and sequester carbon in the soil; they recycle our waste, which exudes heat and greenhouse gases. But their effect on global warming cuts many ways. The methane-producing microbes of the Alaska permafrost were coming alive and posing a toxic threat by thawing, eating carbon, and producing carbon dioxide. Because the Earth is a closed system, in which microbes create most of the gases in the atmosphere, they play a vital role in carbon sequestration and in recycling.

On one hand, microbes exacerbate global warming. The Arctic soil microbes that sequester so much carbon during the winter were becoming more active as the Earth warmed, thus spewing more methane earlier in the year than in the past. Methane, the waste product of soil and gut microbes, is a potent greenhouse gas when not consumed by the methane eaters.

On the other hand, microbes cooled the air. In the ocean, photosynthetic algae and cyanobacteria like *Prochlorococcus* and *Synechococcus* remove some ten billion tons of carbon dioxide from the atmosphere per year. It is on their metabolites that cooling ocean clouds may form. Together, ocean phytoplankton produce up to an astounding 50 percent of the earth's oxygen, and they recycle nitrogen and oxidize ammonia, At the ocean bottom, the University of Delaware researcher Jennifer Biddle studied the microbes that eat the methane at seeps from North Carolina to Antarctica, from Peru to Newfoundland. "We didn't know they ate the methane!" she told me. "The chemists discovered this. This guy I worked with, Chris Martens, saw a flux of gases from the sediment in patterns he could not explain," she said. "He had no explanation so he said, maybe it's the microbes. Chemists love to throw microbes under the bus!"

It was indeed the microbes. When Biddle traveled in Alvin to the bottom of the ocean to understand how, her dives into the depths changed her view of life. "The ocean we see, dolphins and boats, that's not the real ocean," she insisted. As you descend, the vision darkens, and "you see a couple flashes of light that are the bioluminescent markers. I kept looking out the window. It took an hour and a half and it was a test of optimism." She kept hoping something would change. "There's a lot of nothing. On the other hand, that's still something. The water's filled with microbes. They're processing things."

Then, suddenly, she came upon the hot vent, and an explosive world of color and life appeared out of nowhere. "My face was at the window! It was like nothing I had even seen." Worms with no head or mouth, gigantic red shrimp, fish with natural flashlights over their heads. "I'll never look at the world the same way," she recalled.

All this was to understand how microbes play a role in shaping global climate. John Stolz proposed that microbial communities in sediments and stromatolites played a key role in softening global warming. Microbial mats have the diversity of the tropical rainforest, he wrote. They operate as biochemical reactors, burying carbon, as do the diverse communities of purple and green sulfur bacteria of salt lakes such as Mono Lake and the Great Salt Lake. Studying the stromatolites of the Bahamas and Shark Bay, Australia, he observed the complicated communities of cyanobacteria served as carbon sinks. On the other hand, the dangers of bacteria in cattle contributing to methane buildup caused a Canadian company to explore a natural antibiotic that could decrease their production of methane, mainly in the form of cow belches, by some 8–20 percent, and the state of California to move to limit allowable cattle pollution.

Single-celled fungi also helped to shape the weather. In the November 2015 issue of *PLOS One*, researchers at Ohio's Miami University demonstrated that airborne mushroom spores attract droplets of moisture, seeding clouds and eventually falling back to Earth in rain. Nicholas Money, the study's lead author, speculated that this process created a positive feedback loop, in which fungi whose growth is stimulated by rain catapult their spores into the air to encourage more rain. Millions of tons of fungal spores are released into the atmosphere every year, meaning that mushrooms likely affect regional precipitation.

The search for microbe-based biofuels also accelerated. Standard ethanol, made from corn in the United States, sugarcane in Brazil, and sugar beets in France, required a great deal of energy to produce. Sugar cane was by far the most efficient biofuel, and Brazil pulled down thousands of acres of rain forest to grow the sugar cane to make ethanol. Corn ethanol, for its part, probably did little to counter the effects of fossil fuels because it was so inefficient. Companies scoured the Earth for new microbes to break down more common plant cellulose– much harder to metabolize than glucose—and found candidates in such unusual places as volcanic soils (*Sulfolobus*), the hind guts of termites (*Trichonympha*), and the swamps that caused jungle rot in World War II soldiers (*Trichloderma*). All three became the focus of intense interest as fuels.

In agriculture, companies such as Novozymes, Monsanto, and Bayer Crop Sciences were commercializing soil bacteria, and although several start-ups also worked to commercialize microbial cocktails, research had barely begun. The United Nations designated 2015 as the International Year of Soil. Governments, funders, and researchers studied the roles of microbes in achieving food security as population grew and climate change lowered food-crop yields. Until then, only rarely did their initiatives consider the potential of beneficial soil and plant bacteria, adapted through millennia to aid plants in their battle for survival, leading the American Academy of Microbiology to sound a policy call titled, "How Microbes Can Feed the World."

Yeast also became a hit. Using the new gene-editing technology called CRISPR/cas9 (short for clustered regularly interspersed palindromic repeats), companies reengineered their old standby, yeast, to make such commodities as silk, palm oil, and morphine. CRISPR was a much faster, cheaper, more universal, and better automated method of gene editing than older methods. It was also a bacterial invention, with the ability to target genes simply using the cas9 (CRISPR-associated nuclease) enzyme's target. At the Argonne National Laboratory in Chicago, Roser Matamala was the principal investigator of several Department of Energy projects studying the amount of carbon stored by microbes in Alaskan permafrost as researchers worried about the effect of a warming Earth on that critical planet-cooling storage.

Some microbes could devour plastic and others potentially might scavenge carbon. At the University of California, San Diego, researchers devised nanobots that could scour the ocean bottom, eating up the excess carbon that rained down from all our coal and oil refineries. The nanobots used natural enzymes as their fuel sources.

There are five stages, roughly, of a scientific revolution, Thomas Kuhn wrote in his *Structure of Scientific Revolutions*. First was an anomaly, such as those Miller, Wallin, and Margulis noted. Then came a crisis and proposal, such as Woese and Wächtershäuser had offered. Then came rejection in an atmosphere of personality-driven chaos, like the science battle between soup theorists, genetics-first supporters like Jack Szostak, and cell-wall-first proponents. Then came correction, synthesis, and convergence. By 2016, as researchers sought to address multiple crises, the field was still in its adolescence, generating more heat than light. Scientists were forced, like the microbes themselves, to cooperate. The new science was a worldwide social effort, said Slava Epstein, not the product of a lone genius. "We pulled from thirteen disciplines," Dianne Newman said. The joint efforts, McFall-Ngai told me, often led by women like herself, or Newman, Handelsman, Barge, Cox, Chen, Murray, Foster, Visick, Biddle, Sieber, and many others, countered some of the notion of a single crusading scientist who overturns a paradigm.

"First they ignore you," said Gaetan Bogonie, echoing a host of others. "Then they attack you. Then they say, oh, we knew that all along."

All that was needed was money.

The money game

From 2011 to 2015 investment in microbiome-based companies rose by 485 percent, to more than $100 million. In the first ten months of 2016 it was over $600 million. Specialized venture capital firms like Seventure Partners raised hundreds of millions for investing in the new microbiome companies. Pharmaceutical giants like Johnson and Johnson, Abbott Labs, Pfizer, Roche, and Novartis established divisions devoted to the microbiome. Even in such luxury industries as perfume manufacturing, Boston-based Gingko Bioworks called microbes the "future of the fragrance industry," with its plans to market a sweet-smelling genetically modified yeast in a new fragrance. Not only that, it could be manufactured more cheaply, with tighter control, than conventional fragrances. Some shampoos, soaps, and toothpastes began adding probiotics.

The microbiome market was expanding in Europe and expected to grow rapidly in the United States, where appreciation for the cheese and wine fermenters of tradition was less developed. The goal was to produce microbe-based therapeutic and diagnostic treatments for diabetes, Parkinson's, obesity, chronic and autoimmune diseases, cancer, and depression. Many experts linked the rapidly rising rates of depression in women in particular to changes in our natural gut microbes. Pro- and prebiotics, nutraceuticals, and drugs and devices were being developed by a dozen companies worldwide.

In Cambridge, Massachusetts, the company Synlogic was expecting in 2017 to market a capsule with engineered *E. coli* to treat a rare metabolic disease. The businesses involved in synthetic biotics included Amherst, Massachusetts–based Ernest Pharmaceuticals, which was developing engineered salmonella to treat cancer, and San Diego's GenCirq, working on cancer-killing bacteria. The other new companies included Belgium's ActoGenix, France's Enterome Biosciences, Japan's Yakuit, and America's giants DuPont and Merck, as well as start-ups such as Metabolomics, Microbe Therapeutics, Viterra, Vedanta, and Onsel, among others.

The idea of modifying microbes to fight disease was not new. The tuberculosis vaccine and others were created from weakened pathogens. But stock prices of the publicly held companies fell and rose with the new hype because the regulatory climate remained murky and highly challenging. In the United States in 2017, some eight to ten genetically modified microbes were awaiting Food and Drug Administration approval for clinical testing.

One investment company, Flagship, in Palo Alto, California, made a series of strategic inroads into micro-biotech investing, focusing on one of the most advanced companies, the Boston-based SERES Therapeutics, which had three types of microbe or exobiotic-based microbe drugs in development—SER 109, to treat *Clostridium difficile* infections; SER 287, to treat ulcerative colitis; and SER 185, to treat enteric pathogens. But its stock plummeted when one much-anticipated clinical trial failed. Evelo Therapeutics sought to deliver microbes that would improve immunotherapy by triggering the body's immune system to attack cancer-causing tumors, attracting a $35 million Flagship investment to study the microbes around tumors.

The result was a need for a conceptual breakthrough that did not separate microbes into beneficial and harmful types, as those were simply the unintended results of each organism's struggle to survive. Studies showed, for instance, that species interbreeding, as happened with Neanderthals and *Homo sapiens*, was much more common than was thought. We lived in a new uncertainty where the individual ended and the community began. Developed in the right way, microbes may help save the oceans and forests from us. Ignored, they might help drag us down.

The most promising of the microbial discoveries presaged new approaches to some of society's most pressing problems—food shortages, climate change, the energy crisis, failing drugs, and a new approach to global health. Already a number of pro- and prebiotics were being added to foods ranging from fried chicken to frozen yogurt, though it was hard to believe probiotics survived the frying process. Companies raced to develop pro- and prebiotics for crops.

But the potential profit also carried the risk of exploitation. The Archaea enzyme discovered at Yellowstone and used in DNA kits, for instance, was generating $10 million a year for Hoffman–La Roche. The cash-strapped national park never received a penny.

With such a complex challenge, a global government effort was needed, said the British economist Jim O'Neill, a former Goldman Sachs manager and Commercial Secretary to the Treasury. "It would be a classic combination of taxation, government policy, and commercial investment," he said in a special presentation hosted by *Forbes* in April 2016. In Europe a microbiome initiative was announced; in Japan and China a joint agreement was reached.

Toward that end, more researchers were needed who could serve as a bridge from the corporate world to the academic, from pure origin-of-life science to translational medicine that would devise new, targeted treatments for the clinic, from the science conference to the press conference, from the personal to the public. A few were trying.

Love medicine

Jack Gilbert's band Turnaround Time was playing a University of Chicago Medical School reception. "My band is mostly surgeons," the thirty-nine-year-old Gilbert told me on a visit to the university hospital, where he worked with clinicians to study the microbiomes of children from Chicago's South Side, and where he was asked to create a state-of-the-art postoperative lab to prevent surgical infections. "The chief of surgery is my lead guitarist, with an anesthesiologist on bass. I have a heart surgeon on drums. Well, I did; the guy now is a medical resident. I'm lead vocals and guitar," he said, walking rapidly past the residents and nurses as I scrambled to follow.

Gilbert had a main position at the University of Chicago Hospital in the Department of Surgery and a joint appointment as the group leader for Microbial Ecology at Argonne National Laboratory. He used molecular techniques to explore microbial ecology in human and natural settings. He gave TED talks and wrote National Public Radio spots, and he sat on the advisory board of the Genomic Standards Consortium. Listed as one of the world's fifty most influential scientists by *Business Insider*, he had made the news for a discovery that each of us has a distinctive microbial cloud.

Gilbert wore a black T-shirt and talked rapidly about his multiple projects while showing off the illustration of microbes his wife had made for his forthcoming article. On the medical side, he worked on bacteria and diseases such as autism, Alzheimer's, Parkinson's, depression, anxiety, surgical infections, and also obesity and metabolic diseases. In biotechnology, at Argonne he was interested in how bacteria could be used to augment industrial processes, from growing bioengineered crops to making oil easier to refine.

Gilbert was in Nottingham working for Unilever when it offered to send him to Antarctica for eighteen months to study microbes that produced a natural antifreeze. "I was like, somebody is going to send me to Antarctica!" He paused to catch his breath, staring out of the hospital window at the skyline on a bright October day. "I jumped on the bandwagon," he said.

He had a good reason: The London-raised Gilbert had an eight-year-old son, Dylan, who was a high-functioning autistic. At home Dylan read the same books over and over, and he knew every song he and his father played. Now his father was spearheading a global effort to map ecosystems dynamics from the scale of oceans, to that of fields and trees, to that of newly built hospitals. When I asked whether he would try to use his knowledge to change Dylan's microbiome, Gilbert said, "Look, I'm not going to experiment on my son."

With that we were out again on the South Side, outside the hospital smells. The most salient new cancer treatments made use of gut microbes, two studies reported, playing key roles in enhancing or inhibiting the antitumor response of chemotherapy. In a remote acidic hot pool, researchers wanted to tap into acid-loving microbes to deliver insulin drugs to diabetics through the stomach.

By 2030 biofuels could comprise up to 20 percent of marine fuel, a spokesperson for *Hellenic Shipping News Worldwide* reported. In Washington State, Hardwood Fuels released a series of ads on the use of biofuels in petroleum production.

Extreme microbes were set to contribute to new ways of manufacturing, much as soil microbes could be manipulated to increase crop production, a Vanderbilt scientist was discovering. Synthetic tools for analysis and delivery had improved so much that Silicon Valley investors were piling into synthetic biology. The Bill and Melinda Gates Foundation made antibiotic resistance their number-one priority for the coming decade, and the George Soros Foundation followed suit.

The world's smallest organisms may help us solve some of our biggest problems. They worked slowly, but they worked. They also reached the peak of a scientific wave of hype and investment. To understand how that might happen, I made a personal journey, to one of the sites of the revolution's beginning.

14

A River Runs through It

The Hope and Hype of Microbes

Steam rose from the three large, white Yellowstone mounds on a cold late-summer morning as a few straggling tourists fell behind or wandered off to see Old Faithful. I walked with the research team of Eric Boyd, of Montana State University's Department of Microbiology and Immunology. But I was on a separate trek, to an obscure, virtually unknown black pool. The wind whistled around the huge volcanic cones that formed over a fault line, under which bubbling magma heated the water and allowed it to percolate up. Boyd's young team was extracting RNA to see what lived there, seeking to understand how life emerged and what sources of energy might have supported early life on Earth. Coyotes howled in the distance.

Here was where the diversity of the microbial world, of all life, was first revealed. An environment like this might have given rise to life. In one pool in 1994 and 1998, some thirteen new phyla and eighty new species were found by two young people in two short expeditions. How many others were there, in the fourteen thousand other pools, and what capabilities might they have? What could they tell us about the origin of life and the future of the planet? A host of researchers was trying to answer questions like that. It was what Boyd's team was attempting to do.

"It's hot, it's deep, it's hydrothermal, and it's beautiful," Boyd said. Tall and burly, he wore green camping pants and a red parka over a white long-sleeved Yellowstone tourist T-shirt he had grabbed one day when the temperature plummeted. We were miles off-road in the Sentinel Meadows, standing on top of a plateau of twenty-five-foot-high white cone that gurgled ominously. Almost every chemical reaction we need, ancient microbes invented. Like similar hot pools in Kamchatka, Iceland, Germany, Japan, and Australia, those in Yellowstone Park had created a sulfuric acid wonderland.

Thump! The ground shook beneath our feet. Thump! A monstrous hammer shook the white high hill. "Jesus Christ!" Boyd's doctoral student, Melody Lindsay, jumped. She was a musician who sometimes brought a miniature harp to play.

"Feel it?" Boyd asked. "Like it's going to blow?"

Thump! The water rose and boiled in its geyserite cauldron and then, suddenly, it erupted in a white-hot spray, spewing over the twelve-foot-high baked hardpan cone, which resembled something from the surface of Mars. Boyd turned to his group. "You're missing your measurements!" They scrambled into action.

We had snaked down along the Gibbon River, talking nonstop about red and green phototrophic bacteria and the evolutionary history behind their metabolisms; fermentation (since Boyd brewed beer), which involved the acetyl CoA pathways; and the new study of electron bifurcation, a quality of some Yellowstone microbes that may have provided the first metabolic pathway on Earth.

A leader in the next generation after Knoll, Hazen and Newman, Boyd was finding important clues in these ancient, obscure microbes at Yellowstone, Mono Lake, the Great Salt Lake, and the Robertson Glacier in British Columbia. Funded in part by the Department of Energy, his team had recently won two major NASA grants. His group studied hydrogen metabolism, sulfur respiration, the origins of electron transfer, and the varieties of microbes in the world's most famous hot springs. A former football player and pitcher, he was, like me, a fan of the nonfiction writer Norman MacLean. He also knew my desire to go back to the origin of the microbe revolution, Yellowstone's Obsidian Pool.

"Bring your bear spray," the ponytailed Boyd told me as he hoisted his pack. "Let's go."

A new language

We are beset by a rise in incidence of a strange wave of autoimmune maladies. The number of sufferers from asthma and other autoimmune diseases is growing. My children's beloved nanny died of a heart attack exacerbated by her asthma. Many of my children's friends in junior high school had asthma, some form of autoimmune disease, attention-deficit disorder (ADD), or a compulsive disorder. The number of young women suffering from depression was skyrocketing. So too were rates of mild autism. Could our microbes really be contributing to some of these illnesses?

They were responsible for the nearly catastrophic crisis of antibiotic resistance and quite likely had a lot to do with why obesity rates were shooting up. Evidence suggested that our gut microbes played some role in the rising rates of asthma. Researchers at the University of British Columbia had linked the absence in newborns of four types of gut bacteria—*Faecalbacteria*, *Lachnospira*, *Velnollis*, and *Rothia*, a healthful mix they nicknamed FLVR—to an increased risk of asthma. Using data from the Canadian Healthy Infant Longitudinal Development Study, the researchers discovered a "100-day window" after birth when newborns needed to be exposed to their mother's microbiome. It could happen through breastfeeding or vaginal birth. The infants lacking FLVR also had fewer short-chain fatty acids, produced by bacteria in the gut, which enhance an infant's developing immune system. Another finding suggested that children raised on farms had lower rates of asthma than those raised in cities or suburbs.

There were far too many gradations along the autism spectrum to allow doctors to point to a single cause. Still, stomach ailments often accompany the syndrome, and the culprit, according to several studies, could be an absence of *Prevotella*, leading the author Michael Pollan to declare that microbe, in essence, a wonder bug. Researchers at Arizona State University's Biodesign Institute and elsewhere had linked a lack of diversity in the gut microbiome to autism in children. Some were even proposing probiotics as a treatment for young people on the autism spectrum. But the connection was murky.

For obesity, the link that once seemed so simple now looked more questionable. A lack of diversity in the gut microbiome was positively correlated with obesity in several initial studies, and in many studies with incidences of type 1 and type 2 diabetes. However, new data suggested otherwise, at least in humans. The University of Michigan Health System report, which pooled a great deal of human data, found no clear commonality of species in the gut microbiomes of obese people. The more data that was added, the less correlation was found. The NIH-funded study, conducted by the microbiologist Patrick Schloss, used the newest machine-learning big data tools to conclude that "there is really no healthy microbiome," said Schloss.

At the end of 2016 the new company of Caltech's Sarkis Mazmanian, Axial Biotherapeutics, raised some $19 million in seed funding for gut microbiome–targeted therapeutics to treat neurological disorders. At the same moment, researchers in Cork, Ireland, were headlining the Amsterdam-based Mind, Mood, and Microbes conference. The enzymes of some of the heat-loving Archaea of Yellowstone's pools were coming to play an important role in industrial biotechnology, and it was clear microbes played a key role in the cycles of carbon and global warming. Discoveries from life at the Earth's furthest extremes could help to create technologies to reshape the future.

Many factors came together to cause the explosion of understanding new connections, Dianne Newman had told me when I began. The rapid sequencing of DNA, coupled with improvements in computational genetic tools, produced an initial flood of intriguing results. Now the global effort was for big data, requiring many more samples of gut microbiomes, from more diverse populations, ethnicities, and ages. We knew about only 2 percent of our microbial metabolites, their biochemical products, Miller said. Japanese scientists had uncovered ancient microbes some two miles beneath the sea floor, a new record, off the Shimokita Peninsula. From eleven kilometers deep in the Pacific's Marianas Trench, to as deep as we can drill for diamonds in South Africa, to the heights of the stratosphere, microbes made up some 19 percent of the Earth's biomass. "We need," Newman concluded, "a whole new language to understand the coevolution of microbial metabolism with the chemistry of the environment."

Still new things

A front was moving in as we stepped off the path by the Dragon's Lair, a lake-sized cauldron that simmered as it once had when it spewed out and killed hundreds of trees and animals forty years ago. We hiked past the crowds and deep into the most bear-infested region. "I have a permit," Eric Boyd told a man in a Ranger shirt at the Mud Pools viewing platform. "I'm telling you so you don't follow me."

Boyd was hard at work as corecipient of the NASA Rock-Powered Life grant, studying the California Coast Range Observatory, the desert of Oman, and the underwater Atlantis Massif, investigating the molecular biology of microbes in extreme environments to better understand how life emerged and what sources of energy might have supported it early on, exploring the potential of such microbes to provide new types of bioremediation and even energy sources. It's a continuum of geochemical processes generated by the earth's volcanism, gases, and the ebb and flow of water and of carbon dioxide and hydrogen, Boyd was explaining, perhaps close to the metabolic processes of LUCA that Bill Martin studied. "Hydrogen is the key to ancient life," Boyd said.

All I was thinking was that a young tourist had burned to death after falling into a Yellowstone pool, and that a veteran park worker had been mauled by a bear. We disappeared off the path and into the forest. "He didn't have his bear spray," Boyd said of the park worker. I stumbled over singed pine limbs from the 1984 park fire, passing button flowers, phlox, larkspur, and monkshood, into a hilly wood, whistling and clapping all the while. If a grizzly was around, she would know where to find us.

"You have your spray?" Boyd asked me.

I clutched my pants pocket, suddenly remembering that I had thoughtlessly tossed the blue can he had handed me into the back seat. I was also remembering that Yellowstone sat on top of a fifty-mile-wide supervolcano. Tramping through the woods, I whistled again as we clambered over a rise, down into a valley, diverted to avoid a swamp where it appeared some ATV had illegally driven off-road, and then found our way to a clearing where we followed a bison trail, getting soaked to our knees. Lighting flashed, peals of thunder rolled, and rain lashed our faces. "Is that your rain gear?" Boyd asked, dubiously, of my windbreaker and tennis shoes.

When Boyd was growing up in Iowa, his family often took trips out west. As soon he graduated from Urbandale High School, he bought a 1985 Volkswagen Westfalia to drive to Colorado, where he rescued a red Siberian husky and applied to graduate school. He had won a post at Montana State, which meant that he also had the great good fortune to be thirty minutes away from Yellowstone. With his laconic, challenging sarcasm, he became a Twitter star at the recent NASA conference when he said, "I'm only telling you about the experiment that worked, not the one hundred that didn't." Boyd earned a Ph.D. in extreme microbiology at Montana State and was awarded a postdoc as a NASA Astrobiology Fellow. He kept a dusty replica of the Urey-Miller flask by the window of his mountain office.

The smells of sulfur rose as hot water vapor mingled with the chilly late-August air. I clapped some more and stopped. We were in the meadow suddenly, free and clear. The sun peeked out, and Boyd pointed to two plumes of steam rising in the distance, like the smoking remnants of a battle. "There it is," he said.

The hope of microbes

The people in the microbe story, from Pasteur and Wallin to Margulis and McFall-Ngai, from Lyte and Stolz to Boston and Hsiao and Chen, tried hard to unravel a series of highly implausible mysteries. Was there one symbiosis, or many? If microbes exist on other planets, would any place else in the universe have given rise to the same weird combination of complex life forms as Earth, of these plants and animals? What could extreme microbes do for us?

Microbes helped us to metabolize vitamins from vegetables and allowed cattle to eat grass. They gave plants their ability to grow from sunlight and created, with sea microorganisms and plant chloroplasts that began as microbes, all of the Earth's oxygen that we breathe, if one counts chloroplasts as the vestige of microbes they once were. They may help preserve our best emotions, as well as some of our fitness and mental health. Their signaling chemicals gave us most of our medicines. They enable our plants to grow, help our children's brains to develop, and train our bodies to fight disease. Without them, we would not exist as we do now for very long.

Meanwhile, the battles raged about the origin of life, and of complex life. It now appeared that symbiosis, rather than the simple slow mutation of single genes, was a key driver of this evolution. Once, billions of years ago, a microbe swallowed a living bacterium, and that bacterium kept some of its genes, creating the mitochondrial batteries of the complex cell. It seemed that some of the most important genes of plants and animals, those for metabolism and replication, originated in Archaea. The human gut microbiome, with its Archaea, carried relatives of some of these ancient microbes that played a key role in our health and nutrition. None of the Archaea have, as of this writing, been found to be pathogens. Perhaps this understanding could help lead to new medicines. Microbes made energy sources—methane to burn for electricity—from industrial wastewater and acid. They cleaned up oil and radioactive contamination. Some might help to cool the Earth.

With a microbiome numbering in the tens of trillions, no human being is ever alone. Articles and books had made that point. With new tools that can understand the interaction of geology, genes, and behavior, we are on the verge of a revolution that takes advantage of a new understanding of the power of microbes. The most promising of the discoveries offered critical help toward solving some of society's most pressing problems—food, climate change, the energy crisis, failing drugs, and a new approach to global health.

In May 2016 the U.S. government rolled out a long-awaited $121 million National Microbiome Initiative to link researchers in different programs. The next month Bill Gates addressed a packed audience at the American Society for Microbiology meeting in Boston's sleek, cavernous World Trade Center, filled with new company kiosks and pricey public-relations swag. "We are still trying to eradicate communicable disease in the microbe era," said Gates, describing the grants his foundation was making to stamp out malaria in Africa and typhus in India even as we seek out new microbial sources of energy, medicine, and waste remediation. The World Health Organization called microbes "the new frontier of medicine and science."

Perhaps the research would pay for itself, as microbiotech, when the American science-funding climate threatened to tighten or change after the 2016 presidential election. At the microbiology conference, investors wandered the gleaming hallways, past the gigantic, slick marketing floor of companies hawking the new tools of big-data analysis—proteomic, metabolomic, and genomic—promising to study the microbiome, seeking out free gifts to bring to my students in the form of totes, pens, refrigerator magnets, and posters. I could not resist a double amoeba–shaped bright-orange bottle opener. "Productive Pipetting!" proclaimed one sign. A Microbe Art competition featured a Sri Lankan student's portrait of Louis Pasteur painted with *Chromobacterium violaceum*, a gram-negative bacterium. The gram-negative bacterium gets its color from a purple antioxidant that it produces.

I bumped into the Loyola Stritch School of Medicine's Karen Visick, who was working on the bacterial genes that signal biofilm formation on hospital catheters and probes. Then Australia's talkative, bespectacled Phil Hugenholtz, in khakis and horn rim glasses, the second person in Yellowstone to launch the revolution in microbial diversity, paused to chat after his talk. "Susan Barns taught me how to work the sampler," Hugenholtz told me of his trip to Yellowstone in 1998, when he pulled mud from Obsidian Pool. "I found twelve new phyla of bacteria in that one pool. It changed my career." As director of Australia's Centre for Ecogenomics at Queensland University, Hugenholtz was proposing a new tree of life based on whole genomes, rather than the ribosomal RNAs that launched the microbe revolution.

Haunted by waters

At Yellowstone with Eric Boyd, I marveled at the thinness of the crust covering the supervolcano beneath us as we walked. At some points, the upper magma chamber was only four kilometers below the surface. Across the valley, Obsidian Pool lay in an opening of pines and grass before distant hills. This cauldron in the Hayden Valley was where the diversity of the microbial world "jumped out at us," Susan Barns told *Discover* some twenty years ago. The black nine-by-twenty-seven-foot pool resembled a tucked barbell rimmed by shimmering obsidian sand made of volcanic glass. The water was hot and steamy, stinking of rotten eggs, full of methane gas. High in hydrogen, iron, and sulfur, it had white streaks of carbonate and a big, anvil-shaped black headstone where the water pushed up the sand. Here a new understanding emerged: There were many more species of bacteria and Archaea than we imagined, and microbes were more important and fundamental as the diverse underpinning of our environment than all of visible life. "Pretty cool, huh?" Boyd said as we circled the sulfurous, bubbling mud pool.

The amazing thing was that relatives of these same Archaea also lived in our guts, and while some companies were selling biotics to kill them off, Jessica Sieber and others had discovered they nourished the desirable gut-bacterial makers of the short-chain fatty acids (SFCAs) in the colon.

Here was the site of the single most important microbiological discovery of its time, leading to what *Science* called "a new paradigm for understanding life on Earth," and it was linked to what I hoped would be the prime of my life. In this one pool a former art student found thirty-one new species of Archaea, and one Australian found some fifty species of bacteria. Life on Earth was infinitely more diverse than we had imagined. Microorganisms drove and adapted to changes in climate, and in cataclysmic events such as volcanoes, meteorite bombardments, war, and human pollution. "Every metabolic process we depend on they invented," Boyd said, "most of them a really long time ago." They communicate in ways that can trigger a plague or restore an ecosystem. The planet was a vast chemical apparatus, and the unseen world was speaking to us in ways we are only beginning to understand.

And this was only one pool. What lived in the other fourteen thousand pools? Each simmered at a different temperature, with different acidity levels along the entire range of extremes of chemical composition and energy gradients. Some featured the beautiful prismatic cyanobacteria springs that attracted millions, and others had something no tourist valued. What lived in them, deeper than the ten to twenty centimeters where two young people took samples? All we had so far was whatever lived close enough to the surface to be caught in a bucket. Beyond that no one had ever looked.

Much of what we can do, microbes can do better. If one added their viruses, the phages, there were four million varieties of viral proteins about which we know nothing. They were everywhere — in soil, volcanoes, hot springs, Antarctic ice, the stratosphere. Some destroy their host, some infect their host without killing it, some shape its behavior to suit themselves, and some provide critical benefits. Our pace quickened as we marched back through the wild white geraniums, tiny Indian paintbrush, and fireweed. What struck me most was how close benefit and virulence lay, almost at the limits of language or analysis. "Under the rocks are the words," concludes Maclean as the narrator of his life in *A River Runs Through It*. "And some of the words are theirs. I am haunted by waters."

From new tools to discovery

A new understanding, that microbes form the underpinning of our bodies and our environment, had taken hold of much of the scientific community. With tools that can understand the interaction of geology, genes, and behavior, we are in a golden age of exploration that may take advantage of a new understanding of the power of microbes. The most promising of the discoveries offered insights into addressing some of society's most pressing problems — food, climate change, the energy crisis, failing drugs, and a new approach to global health.

At the Weizmann Institute in Israel, researchers discovered that one could formulate the best diet for a person based on that individual's microbiome. In London, a basement factory manned by robots churned out synthetic DNA for manmade cells. In Copenhagen, researchers reversed photosynthesis to make fuel out of dead plants. In Arizona, researchers found dozens of unknown bacterial strains in the films of walls in caves that resembled the environments of Mars. We had reached a stage, said Washington University's Jeff Gordon, where many people alive today would have been dead if they had not received microbiome transplants. But the push today was for far more targeted drugs and microbial remedies than the crude, scorched-Earth fecal transplant.

Gut microbes affected health, mood, and behavior. We have only about a dozen genes in our genome to digest complex carbohydrates, for instance. We could not digest fiber at all. Our gut microbes accomplished that for us. They affected our mood by producing the hormone serotonin directly, by triggering our bodies to produce serotonin, or by triggering the immune system. What we needed most of all, said most researchers, was an international Unified Microbiome Initiative.

But they also provided a way out for remediation and renewable energy, using even very old technologies of breaking down dung to make methane to burn, of cleaning toxic waste, and being engineered to bring us closer to new medicines, like phage viruses to weaken some of the bacteria that beset us. Some researchers suggested fertilizing the oceans with iron, to enhance carbon sequestration. We had already reengineered, through agriculture, vast swaths of the Earth. The focus now was on preserving the Arctic and Antarctic permafrost, where so much carbon was buried. Microbes cooled the earth by consuming millions of tons of carbon and methane and removing hydrogen from landfills and brownfield sites and the great sinks at the ocean's bottom.

A host of companies and researchers turned to next-generation technologies to scale up the discoveries into true sources of renewable energy. In California, Sapphire Energy was on the cusp of marketing crude oil made from algae. Algenol, in cooperation with Dow Chemical, was doing the same for the ethanol, engineering bacteria to enhance the productivity of fermented corn alcohols. In Brazil, several new companies were using the much more efficient biofuel made from sugarcane. At UCLA, the microbiologist James Liao was engineering *E. coli* to make specific alcohols for different kinds of fuel cells.

A new RNA discovery from the Scripps Institute researchers David Hornung and Jerry Joyce claimed to have shown that a ribozyme could link with RNA and create more genetic material—the first time RNA had been shown to replicate itself without enzymes. The RNA in our cells was turning out to be much more complex and important than anticipated, since it modified the actions of genes. RNA now made such an inviting drug target that the garrulous Joyce had joined Novartis Pharmaceuticals, in competition with Alnylam, the other major RNA interference company, and no longer talked as freely as he once had about his enthusiasm for discovering where we came from.

Research teams continued to push back the dates of microbial mats on Earth, beyond the structures discovered by Nora Noffke. University of New South Wales scientists discovered 3.7 billion-year-old microbial mat remains in Greenland. In a remote Hudson Bay outcrop, University College London researchers in 2017 claimed to have found tiny tubes much like those formed by the Yellowstone vent bacteria, dating back anywhere from 3.77 billion years to an astounding 4.2 billion, barely an instant after the Earth took shape.

In San Diego, Irene Chen was working on bacterial phages. In Boston, Laurie Cox was working on microbes and obesity. At Delft University in the Netherlands, researchers had created a microbe paste that would automatically repair cracks in concrete. In Raleigh, North Carolina, Novozymes was uncovering new beneficial bacteria in tidewater soil. Washington University, New York University, and University of Chicago were leaders in investigating the roles of microbes in medical applications. In future years a physical will include a microbiome readout as readily as it does cholesterol level. "Stanford's investing in it, Harvard is investing in it," Jack Gilbert told me. "If you go marine, there are Georgia Tech, and the Woods Hole Marine Biological Lab, and University of Georgia," where the researcher Aron Stubbins discovered that hydrothermal vent microbes offered an enzyme that could potentially help metabolize the manmade carbon in pollution. "The anaerobes were here long before us!" said California's

Mike Cox, pausing between calls and visits to raise money for his company, Anaerobe. All of this brought us to a new potential scenario: the microbe era.

The microbe era

From the hot volcanic pools of Yellowstone, alongside the beautiful blue-green and purple mats of cyanobacteria, came the first glimpse of a revolution in vision. Later we saw some of the same microbes in tropical mangrove swamps, sewage effluents, coral reefs, and in our bodies and realized we know only a small percentage of the microbiosphere.

The trouble with the new microbes was that at first no one could cultivate them in the lab. Then we figured out how to do it. The result was bracing—fragments of new organisms, each stranger then the last. Microbe diversity was far greater than anyone had imagined. In fact, many of the early microbes did not exist as specific species at all. For a billion years they traded genes, clinging to life in tidal pools and borrowing freely from one another's genes for metabolism, energy, and respiration. Symbiosis made for strange bedfellows. The methane-producing Archaea, for instance, often lived with cyanobacteria that did produce oxygen. In a harsh Earth, bombarded by meteors, with few continents and vast acidic oceans, violent tides and severe ice ages, of course the early microbes banded together to survive.

We lived in the modern version of that world, still dominated by microbes that have outlasted us by billions of years, on which all plants and animals depend. The immune system appeared to be designed to "recognize and manage the many complex communities of beneficial microbes" that live inside mammals, and not simply to attack invaders, as medical science has long maintained.

As the Human Microbiome Project progresses, the idea that much of animal biology was missing out on the importance of microbes is becoming more widely accepted. Now people were paying attention.

The understanding that life's diversity was mostly unseen was "vastly transformative," Dianne Newman observed, but is still not fully appreciated. Researchers for years had focused on pathogenic microbes, but now it was seen to be smarter to focus on health than on illness. You could not understand disease without first figuring out the core microbiome of a healthy person.

Animals had their origin in the union of microbes. They shared microbial gut communities in much the same way the original microbes shared genes. The whole concept of a tree of life mapping separate species was becoming outdated. It was more of a partnership among species and with the environment that shaped the living world. Microbes played roles in brain development, behavior, the immune system, mood, and our sleep cycles. Pathogens were a normal part of the body. It could be that disease was a matter of changing the balance of a commensal community. After focusing so much on pathogenesis, researchers now saw that virulence and toxins were part of the normal conversation a microbe had with its host animal. They were present all the time. Perhaps disease depended on the signals' context and nuance.

The Earth and our bodies alike exist as a multilayered set of niched communities. Microbes were involved in and affected by many biogeochemical cycles and processes. They helped to seed the clouds for rain and cool the Earth when it became overheated. What is the genetic capability of microbes that could feed on sulfur, arsenic, and heavy metals, and that lived in most every part of the human body?

The ideas had started small—in the ocean shallows, in vents, and in pools like Obsidian Pool. But the idea that people had ignored or disparaged, the potential of microbes to help or hinder our health, had arrived. Our fears, hopes, and understandings are connected with the inhabitants of our bodies and the unkempt worlds we live in.

Each together could hold deep implications for the future. The world's smallest organisms may help us solve some of our biggest problems. This could be the beginning of a microbiome era.

Epilogue

The Croatian island shore featured an array of summer bodies, old and young, skinny and fat, waiting to scramble into the shower, the sauna, and the pool. Our bodies are gardens for our fellow-travelers the microbes. With new tools that can understand the interaction of geology, microbial genes, and behavior, we may be on a precipice, as Duluth's Jessica Sieber put it, of a microbial revolution in understanding.

My family rides a small fishing boat under a gathering storm off the shore of the Pakleni Islands, near Hvar. The shore stretches in rock and beaches, with scrub pines once harvested for their resin, used to seal ships. Snorkeling for hours and looking for octopus, I think of the microbial sedimentary structures present since life began virtually at Earth's first moment. For two billion years ancient microbes thrived without oxygen. Oxygen, created first by cyanobacteria, then by plants, was in the beginning a global toxic catastrophe. From that point, however, complex life could evolve, and the Earth's dominant microbes retreated to the swamps, rocks, and deep-sea vents, and to the bodies of animals and humans. We thought of those hidden microbes as pathogens, but they invented most every reaction that keeps us alive.

At a veranda of a neat, artistically designed farm table where we pause for lunch, I watch the rows of grapevines, thinking of the quirks of microbial metabolism that will shape the wine they make. The hygiene hypothesis of my father-in-law could, it turns out, explain at least part of the reason for the soaring rates of asthma and anxiety disorders, which probably have at least some relation to the human gut microbiome. Researchers around the world applied next-generation metagenomics tools to analyze cells for their full genomes rather than the tiny 16sRNA technology employed by pioneers of the past. Much of what we call virulence might be straightforward signaling among microbe communities. To combat their fast-evolving resistance to our medicines, researchers followed several paths: searching in new soils, reengineering old drugs, and looking at diet and the benefits we derive from staying dirty. The antibiotic-resistance crisis forced new connections among such disparate disciplines as medicine, geology, and chemistry that are changing our vision of the world and the roles of microbes in it.

Some are key to oil remediation and can clean up acid mine and heavy metal spills. They can even implant uranium in the ground and keep it out of runoff into fragile western rivers like the Colorado. They cool the earth by recycling carbon, nitrogen, and methane. They offer sources of energy as well as petroleum remediation, agricultural pest control, and sulfate reduction.

They instilled a new interdisciplinary surge among researchers, geologists, microbiologists, chemists, medical doctors, nutritionists, biologists, marine researchers, and bioinformaticians, using social media and metagenomics tools as well as remote sensors to generate microbe breakthroughs on which new companies sought to capitalize. Slava Epstein's new remote sensing tool, Gulliver, he predicted, would spur a new microbiology without microbiologists. The new gene-editing tool called CRISPR could be operated almost by anybody, proclaimed a new issue of *Science*.

The microbe revolution had taken hold from the roots up, pioneered by interdisciplinary scientists who revealed the power of Earth's microorganisms acting together. Microbes inhabited this planet long before we were here. They created alliances and waged wars on a level more complicated than anything in the Game of Thrones. The unexplored microbes of freshwater lakes, deep mines, the earth's surface, ocean-floor vents, and tidal estuaries, critical to global health, are still mostly uncharted. They also tempted us to overstate their influence. "It's like we have walked to work every day and passed a skyscraper but didn't see it," said Jennifer Biddle. "And now we see skyscrapers everywhere."

We had good candidates for the places where microbial life appeared, and how. Researchers were designing probes to search for them beyond Earth. Some were uncovering ways this hidden world could provide new energy sources, as cleaners of spills and waste.

For all the excitement about harnessing the abilities of the planet's microbes, the proposed Trump administration 2018 budget cuts in science funding were challenging. The proposed cuts ranged from 17–20% of the budgets for the National Institutes of Health and US Department of Agriculture, with a 5.6% proposed cut for the Department of Energy, all of which had funded early microbial research. NASA's funding was held close to 2017 levels, but with sharply reduced outlays for space technology to monitor planet Earth. Some of the cuts would likely be ameliorated by Congressional wrangling and grants from some private foundations, like the Gates Foundation, which was promising expanded support of research into the benefits provided by microbes, but US funding for microbiome research threatened to lag behind the rates of support of some European and Asian governments.

The Earth's metabolic network was created by a microbial and geochemical world that is so interconnected as to defy the concept of individual species. A community of microbes in a swamp or beachhead like this is an integrated construct "almost mathematical in its design," Carl Woese told me. "It is what keeps the biosphere stable, what feeds it, and it shapes our geology and environment."

We were still just beginning to understand these connections among the microbes and the world. The discoveries had forged a new vision of life and its potential. This field brought science into dialogue with an emerging complex and beautiful understanding of relationships between the tiniest life forms and the planet. I felt a new sense of hope. Microbes were not the only things that lived in social groups. Scientists too, and the writers following them, were learning some of the amazing insights of a new approach to medicine, the environment, emotional health, and where life came from and what brought it into being from top experts.

As I turn sixty, I turn my mind more and more to the implications of the microbe discoveries. The chemistry in the universe exactly fits the specifications to create life. We are part of a continuum with the Earth's geochemistry, its ocean cycles, volcanoes, climate, and natural products. We have the responsibility of stewardship of its fragile balance. Each wonder of its creation adds to my awe in its contemplation. That was my particular sensitivity, and what was beautiful to me.

Microbes help us to understand how life functions on the planet, how to cope with human stressing of it, and how to keep that stress from going out control. Most every move they make is a secret wonder, and now many of those moves are no longer secret.

Some Microbes of the Human Body

The bacteria of our armpits include *Propionibacterium acnes*, found on many skin areas; it releases fatty acids onto the skin, which help create an environment that fights infections. *Malassezia* is a fungal microbe that is naturally found on human skin and feeds on lipids; this means it is more likely to be found in places with sebaceous glands, and it helps contribute to healthy skin. *Staphylococcus hominis* for instance, has been identified as one of the culprits responsible for smelly armpits: the bacteria feed on moisture and lipids found naturally in moist areas of the body and convert them into thioalcohols, which have a strong odor.

Acinetobacter — This bacterium is found on skin and in armpits and functions as a protectant from allergens on the skin's surface. It can survive in moist and dry environments.

Brevibacterium linens — This bacterium is commensal and found commonly on skin. It is best known for causing foot odor. It is also involved in fermenting such cheeses as Munster.

Candida — A group of common yeast species that are usually commensal and are commonly found on skin and in the stomach.

Corynebacterium — This bacterium is one of the most common found on skin in general and is always present on the armpit. It is also blamed for armpit odor.

Debaryomyces — This group of fungus species is thought to inhabit the skin. The species group includes yeasts that are found in hypersaline environments, as well as all types of cheese. Harmless to humans, it is thought to be commensal.

Demodex folliculorum — These are microscopic arthropods (mites) that are normally found on human skin. They feed on sebum, are more prevalent during puberty, and are often found on the face.

Malassezia — This fungus is frequently found on the skin and feeds on lipids. It is part of a healthy skin microbiome but in large quantities can be responsible for skin irritations and dandruff.

Micrococcus luteus — This species processes lipids to cause odor and can survive in salty environments. It is a commensal bacterium that is also found in animals and dairy products.

Staphylococcus — This bacterium is found in many parts of the body; it can be pathogenic and cause infections when present in large numbers.

However, it produces fatty acids, which inhibit the growth of fungi and yeast.

Staphylococcus epidermidis — This common commensal bacterium has recently shown potential to inhibit skin pathogens. It can cause staph infections under certain circumstances but is also one of the most common types of bacteria throughout the human body.

Some Bacteria Normally Found in the Human Stomach or Intestines

Bacteroides fragilis — This group of bacteria is generally beneficial to the human body but can also cause infection if any make their way into the bloodstream. They process complex sugars in our intestines so we don't have to. They generate simplified carbohydrates, amino acids, and vitamins.

Bacteroides thetaiotaomicron — This bacterium helps break down complex sugars and carbohydrates. It can also bind starches, which allows humans to use a wider variety of dietary carbohydrates.

Bifidobacterium bifidum — This is one of the most common bacteria found in the human gut. It helps break down long chains of simple sugars, and increased amounts of this bacterium will boost the immune system. It is one of the most well-known ingredients in probiotics.

Enterococcus faecalis — This bacterium has a commensal relationship with the human gut and ferments glucose. Although it is harmless in the stomach or intestines, it can cause urinary tract infections, meningitis, or other infections.

Escherichia coli — This bacterium is well-known as *E. coli*, which can cause food poisoning. However, other strains of this bacterium live in the human gut and provide beneficial services such as production of vitamin K and prevention of the growth of pathogenic bacteria.

Faecalibacterium prausnitzii — About 5 percent of all bacteria in the intestines are *Faecalibacteria*. Without proper levels of this bacterium, a person may be more susceptible to Crohn's disease, obesity, and asthma.

Lactobacillus acidophilus—This bacterium is present in the human digestive tract and is thought to help lactose-intolerant people process
dairy.

Methanobrevibacter smithii—This is the main archaeon in the human gut
and plays an important part in digesting complex sugars by consuming the end products of bacterial fermentation, which contain methane. Too much methane can lead to digestive problems such as consti-
pation.

Methanosphaera stadtmanae—This is the second most common archaeon
in the human gut and has the most restricted energy metabolism of
all known methanogenic Archaea. This means that it is dependent on
acetate, which is abundant in the digestive tract.

Streptococcus—This bacterium belongs to the order of lactic-acid bacteria
and has a commensal relationship with humans. It is one of the only
bacteria that can survive in the acidic environment of the stomach.
This is the primary bacterium involved in plaque formation and initiation of dental caries.

Some Tips for a Healthier Microbiome

Smile (the air kills the anaerobes in your mouth)
Eat well, including fiber and microbial-accessible carbohydrates (such
as yogurts, cheeses, fruits, leafy vegetables, pickles, sauerkraut)
Get enough rest and exercise
Work in the garden
Travel a little less (jet lag decreases microbiome diversity)
Work out your immune system (it's okay to get sick sometimes)
Bathe a little less frequently
Keep a pet
Visit a farm
Avoid fast food
Consider a compost heap

Timeline

1926: Konstantin Merezhkovsky proposes that plant chloroplasts were once freestanding bacteria

1927: Ivan Wallin, University of Colorado Medical School, suggests mitochondria are captured bacteria within the cell

1953: Stanley Miller, "A Production of Amino Acids under Possible Primitive Earth Conditions"

1967: Lynn Margulis, "On the Origin of Mitosing Cells"

1977: Woese and Fox PNAS Paper on the Archaea, a putative
third form of life

1977–78: Jack Corliss and Alvin discover teeming life at hot
vents

1983: RNA splicing is discovered by Tom Cech and Sydney
Altman.

1988: The pyrite hypothesis is proposed by Günter Wächtershäuser

1994: Susan Barns's Obsidian Pool discoveries published in
PNAS

1998: Phil Hughenholtz discovers fifty-four new species of bacteria and Archaea in Obsidian Pool

1999: Texas Tech's Mark Lyte suggests probiotic bacteria could be tailored to treat psychological diseases.

2000: NASA Stardust Mission—comets contain a mix of high- and low-temperature materials

2003: Discovery of alkaline vents, Lost City

2006: Jack Szostak's lab discovers that RNA evolves into more and more efficient RNA

2007: NIH's Human Microbiome Project begins

2008: Robert Hazen makes amino acids at model deep-sea black smokers in lab.

2009: Paul H. Patterson, Caltech: role of microbes in neurodevelopment; Sarkis Mazmanian, Caltech: how microbes modulate neuroactive molecules

2010: Sage College's Susan Jenks and Dorothy Matthews find mice fed a form of *vaccae* from soil had improved cognitive function; Swedish scientists claim that the gut microbiome regulates brain development

2010: Samantha Joye: petroleum-eating microbes clean up Deepwater Horizon spill; Smith and Venter create a synthetic form of life by replacing the genome of an existing bacterium with a manmade sequence stitched of synthesized DNA

2011: NYU's Langone Medical Center Director Laurie Cox, Martin Blaser: the microbiome plays a key role in the fact that obesity in children has gone up some 450 percent since 1996. Ancient relationships perturbed: Tanya Yatsunenko compares U.S. gut microbiomes with those of Malawians in Africa.

2011: Tullis Onstott: thriving microbe in deep South African mine, *Hadesarchaea*, also lives in tidal estuaries and hydrothermal hot pools

2013: Sandra Pizzarello: new organics and water in Sutter's Mill meteor,

2013: McMaster University, Ontario: gut microbes may affect mood

2013: Irish researchers at the University of Cork discover "melancholy microbes" in the gut that are linked with depression.

2013: Synthetic Genomics creates synthetic flu vaccine

2014: Martin Blaser and Laurie Cox discover that if antibiotics were limited for four weeks at the beginning of a mouse's life, gene expression in mice intestines improved; in Norway, gut microbes of fifty-five people and certain bacteria are linked to depression

2014: Loki discovered at "castle" below Norway North Sea, bridge of
Archaea to animals

2014: New class of antibiotic from Maine, teixobactin, discovered by
American and German researchers. Novobiotic, first new antibiotic in
twenty years, targets resistant staph and *Clostridium difficile*, the cause
of contagious colitis.

2014: NASA Curiosity team: methane detected in Mars's Gale Crater

2015: Old Dominion's Nora Noffke takes a second look at Mars Curiosity's
photos; more than 70,000 microbe gene sequences analyzed, the U.S.
Department of Energy expanded microbial genome sequencing for
other industrial and energy uses.

May 2015: Spang et al., "Complex Archaea That Bridge the Gap" (*Nature*),
find that the Thor genome suggests all complex life is an offshoot of
Archaea

August 2015: Philae Lander finds multiple amino acids on Churyumov-
Gerasimenko

December 2015: Briny water on Mars

March 2016: Synthetics Genomics creates first minimal cell

April 2016: Pluto has a liquid ocean

May 2016: Obama White House initiates $121 million microbe initiative

July 2016: Bill Martin traces last universal common ancestor to alkaline
vents

November 2016: NASA Mars reconnaissance orbiter detects buried ice
field the size of Lake Superior

February 2017: Seven Earth-sized planets detected around nearby red
dwarf star TRAPPIST-1

Remote Probes

In 2014, NASA's Curiosity rover detected signs of methane and organic by-products in a Martian crater. Some two dozen other NASA, European, Chinese, and Russian probes have flown or are slated to explore the solar system's life-friendly planets and moons. Fans follow them via Twitter and Instagram.

Moon

2008: NASA Lunar probe deliberately crashed near South Pole to reveal some 6 percent water ice

2010: NASA Lunar probe reanalysis suggests that some four billion years ago the moon may have had liquid water one meter deep

2017: SpaceX announces it will send two people to orbit the moon in 2018

Mars

2004–10: NASA Opportunity and Reconnaissance Orbiter sends back pioneering images and maps of the basins of seeming ancient sea beds, floodplains, and windswept arroyos.

2008–11: NASA Phoenix collected soil and searched for water near the northern pole

2011: Chinese National Space Agency's (CNSA) Yinghua 1 failed to escape Earth orbit

2012–16: NASA Curiosity and Mars Orbiter: Thirty-kilometer journey through Gale Crater and Yellowknife Bay en route to climb Mount Sharp. Strong evidence of dry–wet cycle conducive to life.

2014: NASA Maven; launched in September to study Martian atmosphere

2016–18: European Space Agency (ESA) and Russian (RKA) ExoMars orbiter and Schiaparelli rover

2016–19: Indian Mars Orbiter Mission (MOM) analyzes methane in Mars atmosphere

2019: NASA and ESA planned sample return mission; rover is named Urey, for University of Chicago's Harold Urey

2020: NASA next-generation lander is planned

2020: Chinese Space Agency (CASC) planned mission including an orbiter, lander and Chinese Mars Mission

2020: Space X Mars lander anticipated

2021: Planned Saudi Arabian Space Agency Mars mission

2020s: Mars-Grunt Russian and Finnish mission to collect Mars soil samples

2028: Space X plans to send humans to Mars

2030s: NASA plan to put humans in orbit

Asteroids and Meteors

Churyumov-Gerasimenko

2014–16: European Space Agency Rosetta probe Philae makes an unprecedented landing on comet Churyumov-Gerasimenko and discovers many previously unseen amino acids in the comet's soil.

Ceres (largest object in the asteroid belt,
half again larger than the Earth's moon)
2015–18: NASA Dawn conducts spectrometry and takes pictures and radio
  astronomy readings

Jupiter Moons

Ganymede
2000s: NASA Cassini and ESA Huygens flybys of Jupiter and Ganymede
  take pictures

Europa
2002: NASA Galileo pictures suggest a frozen ocean overlying volcanic
  plates
2013: NASA Hubble photos detect evidence of claylike minerals in icy crust
  and icy water-vapor plumes similar to those detected on Enceladus
2020s: NASA flyby probe is planned

Saturn Moons

Enceladus (closest moon of Saturn, subject to tidal heating)
2010: NASA: reanalysis of Cassini data suggests a large south polar subsurface liquid ocean potentially five miles deep
2030s: NASA Life Investigation for Enceladus (LIFE) probe to land or
  sample plumes is in the planning stage

Titan
2005: ESA Huygens, deployed by Cassini, is first probe to land on moon of
  another planet
2014: NASA Hubble analysis suggests methane rainfall from an atmosphere of material from the Oort Cloud, and a sub-ice ocean saltier
  than the Dead Sea
2029: ESA and NASA planned orbiter and lander under study

Acknowledgments

I would like to acknowledge the roughly 150 scientists and graduate and undergraduate researchers who gave freely of their time and knowledge to my repeated questions, especially those who let me visit their labs and interview them at conferences. Two scientists who read the manuscript in its early stage, Bernard Selling and Eric Boyd, deserve special mention and thanks. Others who spoke to me many times include Nora Noffke, Andrew Knoll, Alison Murray, Jack Gilbert, Sandra Pizzarello, Steven Benner, Bill Martin, John Sutherland, Mike Russell, Margaret McFall-Ngai, Laura Cox, Linda Kah, Robert Hazen, John Stolz, Slava Epstein, Itay Budin, Melody Lindsay, Chris McKay, Barbara Golden, Brenda Bass, Dianne Newman, and many more.

Several excellent research assistants assisted me in the

writing of this book. First was Erica Dix, who conducted background research, wrote up researcher biographies, and helped to transcribe tapes as well as review the manuscript. A second talented research assistant who worked with me from the beginning was Alexandra Gerard. I acknowledge the reporting help of Alexandra Nates-Perez and the thoughtful assistance of award-winning high school biology teacher and current Hollywood screenwriter Bryan Kett.

For their vision and enthusiasm I thank my agent Ellen Levine, the vice president of Trident Media Group, and Christie Henry, the editorial director of Sciences and Social Sciences at the University of Chicago Press. Their expertise and support were critical throughout the writing of this book. I acknowledge also the contributions of my copy editor, Barbara Norton, and indexer Julie Shawvan, who helped to bring *Planet of Microbes* from manuscript draft to finished book.

For her loving support and discerning eye, I thank my wife, Maja.

These people and many others helped me to bring a circuitous but usually energetic quest to this conclusion. In thanking them, I note that any errors in the final book are not theirs, but my own. Finally, I acknowledge the readers of this book, who are now a part of its story.

Notes

Introduction

a homestead's crumbled foundation, too tired   Nora Noffke, phone interview with the author, June 25, 2014. Unless otherwise noted, subsequent quotes of Noffke are from this or another interview.

pulled out a flask of warm water   Daniel Christian, phone interview with the author, January 30, 2015. Unless otherwise noted, subsequent quotes of Christian are also from this interview.

"They're rolled up even"   Nora Noffke, phone interview with the author, May 11, 2015. Unless otherwise noted, subsequent quotes of Noffke are from this or another interview.

some thirty to forty trillion   Ron Sender, Shai Fuchs, and Ron Milo, "Revised Estimates for Number of Human and Bacteria Cells in the Body," *bioRxiv.org*, accessed July 5, 2016, http://bioxriv.org/content/early/2016/01/06/036103.

microbes do most of the clean-up   American Society for Microbiology, *Microbes and Oil Spills: FAQ* (Washington, DC: American Academy of Microbiology, 2011), 1–16.

Several books have covered   Ed Yong, *I Contain Multitudes: The Microbes Within Us and a Grander Vision of Life* (New York: HarperCollins, 2016), for example, focused on microbes in humans and animals. Sean Carroll, *The Big Picture: On the Origins of Life, Meaning, and the Universe Itself* (New York: Dutton, 2016), takes a philosophical approach to the meaning of recent discoveries.

Chapter One

Tesla coil   Jeffrey Bada, "The Origin of Life," accessed July 11, 2016, www .thenakedscientists.com/HTML/interviews/1021/.

his father died suddenly in 1946   Jeffrey L. Bada and Antonio Lazcano, "A Biographical Memoir," *National Academy of Sciences* (2012): 3, accessed July 11, 2016, http://www.nasonline.org/publications/biographical-memoirs /memoir-pdfs/miller-stanley.pdf.

inspired Watson and Crick   Erwin Schrödinger, *What Is Life?* (London: Cambridge University Press, 2012).

"time-consuming, messy"   Miller, quoted in Jeffery L. Bada and Antonio Lazcano, "Stanley Miller (1930–2007): Reflections and Remembrances," *Origin of Life Evolutionary Biosphere* 38 (2008): 375.

One day he attended a campus lecture   Johnjoe McFadden and Jim Al-Khalili, *Life on the Edge: The Coming of Age of Quantum Biology* (New York: Crown, 2016), 272.

"It is possible to create an experiment"   Ibid., 273.

meet Heisenberg and Einstein   American Institute of Physics, Harold Urey, interview with John C. Heilbron, accessed July 8, 2016, https://www.aip .org/history-programs/niels-bohr-library/oral-histories/4927-1.

"Water is boiled in the flask"   Stanley Miller, "Production of Amino Acids under Possible Primitive Earth Conditions," *Science* 117, no. 3046 (May 1953): 528–29, accessed September 6, 2016, http://science.sciencemag .org/content/117/3046/528.

"cloudiness and turbidity"   The Miller-Urey Experiment, accessed July 14, 2016, http://thelivingcosmos.com/The OriginofLifeOnEarth/Miller UreyExperiment_12May06.html.

Glycine is a building block of proteins   Accessed July 7, 2016, https://examine .com/supplements/glycine/.

The article appeared two months later   Miller, "Production of Amino Acids," 529.

"If God did not"   quoted in Robert Hazen, *The Story of Earth: The First 4.5 Billion Years, from Stardust to Living Planet* (New York: Penguin, 2012), 134.

the lakes, foothills, and steppes of Kazan province   Jan Sapp, Francisco Carrapiço, and Mikhail Zolotonosov, "Symbiogenesis: The Hidden Face of Constantin Merezhkowsky," *History and Philosophy of the Life Sciences* 24, nos. 3–4 (February 2002): 413–40.

You could see them in a microscope   Lynn Margulis and Dorion Sagan, *Acquiring Genomes: A Theory of the Origins of Species* (New York: Basic Books, 2002), 14.

and then in a book   Konstantin S. Merezhkovsky, *Theory of Two Plasms as the Basis of Symbiogenesis: A New Study or the Origins of Organisms* [in Russian] (Kazan: Publishing Office of the Imperial Kazan University, 1909).

annual Christmas glogg party   Cold Spring Harbor Laboratory, "Biography 30: Ivan Emanuel Wallin (1883–1969)," *DNA Learning Center*, accessed May 10, 2016, http://www.dnaftb.org/30/bio.html.

"A principle that is so revolutionary"   Ivan E. Wallin, *Symbionticism and the Origin of Species* (Baltimore, MD: Williams & Wilkins, 1927), vii

"Mitochondria Man"   Cold Spring Harbor Laboratory, "Biography 30."

"Stanley Miller inspired"   Robert Hazen, phone interview with the author, August 9, 2016. Unless otherwise noted, subsequent quotes of Hazen are from this or another interview.

Chapter Two

she got in   "Lynn Margulis on Her Life, Symbiogenesis, and Gaia Theory," interview with Jay Tischfield, Rutgers University, accessed July 13, 2016, https://www.youtube.com/watch?v=KlhW12dGfFk.

Margulis challenged him   Dorion Sagan, Skype interview with the author, June 28, 2016. Unless otherwise noted, subsequent quotes of Sagan are from this interview.

"planetary patina"   Lynn Margulis and Dorion Sagan, *Microcosmos: Four Billion Years of Evolution from Our Microbial Ancestors* (New York: Summit Books, 1986), 210.

custody of their two sons   Lynn Sagan, "On the Origin of Mitosing Cells," *Journal of Theoretical Biology* 14 (1967): 274–75, accessed July 12, 2016, http://web.gps.caltech.edu/classes/ge246/endosymbiotictheory_marguli.pdf; Dorion Sagan, interview, June 28, 2016.

"craving after knowledge"   Anton van Leeuwenhoek, letter to University of Louvain rectors, June 12, 1716, quoted in Charles Edward Winslow, *The*

Conquest of Epidemic Disease: A Chapter in the History of Ideas (Madison: University of Wisconsin Press, 1980), 156.

In a book Lynn Margulis, *Origin of Eukaryotic Cells: Evidence and Research Implications of the Origin and Evolution of Microbial, Plant, and Animal Cells on the Precambrian Earth* (New Haven, CT: Yale University Press, 1970).

"All visible organisms" Interview with Dick Teresi, *Discover*, accessed April 16, 2016, http://discovermagazine.com/2011/apr/16-interview-lynn -margulis-not-controversial-right.

should be abandoned Margulis and Sagan, *Acquiring Genomes*, 16.

"it is a process" Lynn Margulis and Richard Guerrero, "Two Plus Three Equal One: Individuals Emerge from Bacterial Communities," in *Gaia 2: Emergence; The New Science of Becoming*, ed. William Irwin Thompson (New York: Lindisfarne Books, 1991).

"Attila the hen" Dorion Sagan, ed., *Lynn Margulis: The Life and Legacy of a Scientific Rebel* (White River Junction, VT: Chelsea Green, 2012), 30.

"biology's great ironies" Michael W. Gray, "Rickettsia, Typhus and the Mitochondrial Connection," *Nature* 396, nos. 109–10 (November 12, 1998), accessed February 17, 2017, http://www.nature.com/nature/journal /v396/n6707/full/396109a0.html.

much the same way James Lovelock, *Homage to Gaia* (New York: Columbia University Press, 2007), 47.

"three-billion-year-old septic tank" Sagan, *Lynn Margulis*, 42.

"on the order of Kuhn" Ibid., 38.

"a cybernetic system with homeostatic tendencies" James E. Lovelock, "Geophysiology: The Science of Gaia," *Reviews of Geophysics* 17 (May 11, 1989): 215–22.

"evil religion" Quoted in Jeff Goodell, "James Lovelock, the Prophet," *Rolling Stone*, November 1, 2007, accessed July 13, 2016, http://www.rolling stone.com/politics/news/james-lovelock-the-prophet-20071101.

"The editorial board must be senile" Sagan, *Lynn Margulis*, 40.

"Life can flourish only" Lovelock, "Gaia as Seen Through the Atmosphere," in P. Westbroek and E. W. deJong, *Biomineralization and Geological Perspectives: Papers present in the Fourth Annual International Symposium on Biomineralization* (Amsterdam: D. Reidel, 1983), 15.

"Gaia is a tough bitch" Lynn Margulis, accessed June 28, 2016, https://www .edge.org/conversation/lynn_margulis-lynn-margulis-1938-2011-gaia-is -a-tough-bitch.

"unbelievable experience . . . there was no box" John Stolz, Skype interview with the author, December 9, 2015. Unless otherwise noted, subsequent quotes of Stolz are from this interview.

"whose contribution should never be forgotten!"   Margaret McFall-Ngai, phone interview with the author, March 21, 2015. Unless otherwise noted, subsequent quotes of McFall-Ngai are from this or another interview.

9/11 was a right wing conspiracy   Margulis, accessed July 14, 2016, https://www.youtube.com/watch?v=g-GFBEX5bjY.

combating global warming   Goodell, "James Lovelock, the Prophet."

many disagreed   Peter Ward, *The Medea Hypothesis: Is Life on Earth Ultimately Self-Destructive?* (Princeton, NJ: Princeton University Press, 2009).

"two previously unconnected facts"   Bill Martin, phone interview with the author, August 10, 2016. Unless otherwise noted, subsequent quotes of Martin are from this or another interview.

"you could understand how"   Carl Woese, phone interview with the author, July 14, 1999. Unless otherwise noted, subsequent quotes of Woese are from this interview.

Chapter Three

"ski really fast" and "let loose"   Accessed March 12, 2017, http://www.the-scientist.com/?articles.view/articleNo/14914/title/Tom-Cech/.

Iowa River Valley   "Thomas R. Cech, PhD," Howard Hughes Medical Institute, accessed June 23, 2016, http://www.hhmi.org/scientists/thomas-r-cech.

doing the experiment himself   Tom Cech, accessed August 18, 2016, http://www.ibiology.org/ibiomagazine/issue-1/tom-cech-discovering-ribozymes.html.

could do the work of enzymes   Accessed January 27, 2017, http://www.hhmi.org/scientists/thomas-r-cech.

later, the songwriter Leonard Cohen   "Sidney Altman—Biographical," accessed June 23, 2016, http://www.nobelprize.org/nobel_prizes/chemistry/laureates/1989/altman-bio.html.

"to fight for what you believe"   Sydney Altman, phone interview with the author, May 15, 2015. Unless otherwise noted, subsequent quotes of Altman are from this interview.

In 1962 Altman read the Nature *paper*   Ibid.

"Then that's what nature is telling you," Meselson said   Ibid.

Today Altman theorizes   Ibid.

Bored as an attorney   Günter Wächtershäuser, phone interview with the author, June 20, 2014. Unless otherwise noted, subsequent quotes of Wächtershäuser are from this interview.

"held sway over science for hundreds of years"   Günter Wächtershäuser, "The

Uses of Karl Popper," *Royal Institute of Philosophy Supplement* 39 (1995): 180.

"*such matter would be instantly devoured*" Francis Darwin, ed., *The Life and Letters of Charles Darwin, Including an Autobiographical Chapter*, vol. 3 (London: John Murray, 1887), accessed June 24, 2016, http://darwin-online.org.uk/content/frameset?keywords=be%20matter%20devoured%20instantly%20would%20such&pageseq=30&itemID=F1452.3&viewtype=text.

"*turned the entire concept of life's beginning upside down*" David Deamer, *First Light: Discovering the Connections between Stars, Cells, and How Life Began* (Berkeley and Los Angeles: University of California Press, 2011), 71.

discovery of springs in Kamchatka Ibid.

"*a process of self-liberation*" Ibid.

the next step had been taken Michael Hagmann, "Between a Rock and a Hard Place," *Science* 296 (15 March 2002): 2006–7, accessed April 7, 2012, http://science.sciencemag.org/content/295/5562/2006.

Chapter Four

She had wanted to get involved Susan Barns, phone interview with the author, February 4, 2000. Unless otherwise noted, subsequent quotes of Barns are from this interview.

"*He worshipped truth*" Mitchell Sogin, phone interview with the author, December 1, 1999. Unless otherwise noted, subsequent quotes of Sogin are from this interview.

"*understanding the origins of life*" Virginia Morrell, "Microbial Biology: Microbiology's Scarred Revolutionary," *Science* 276, no. 5313 (May 2, 1997): 701.

It was a heroic effort George Fox, interview with the author, Astrobiology Science Conference, Chicago, June 12, 2015. Unless otherwise noted, subsequent quotes of Fox are from this interview.

Much of the world's total biomass Woese, interview, July 14, 1999.

"*keeps the biosphere stable*" Ibid.

and named it Thermus aquaticus Shauna Stephenson, "*Thermus Aquaticus*," accessed June 28, 2016, http://www.wyomingnews.com/things_to_do/thermus-aquaticus/article_dd046f75-0178-5793-89f5-8711077e9303.html.

even at the boiling point Thomas D. Brock, "The Value of Basic Research: Discovery of *Thermus aquaticus* and Other Extreme Thermophiles," *Genetics* 146, no. 4 (August 1, 1997): 1207–10, accessed August 18, 2016, http://

www.ncbi.nlm.nih.gov/pmc/articles/PMC1208068/pdf/ge14641207
.pdf.

from its kits   Ted Anton, *Bold Science: Seven Scientists Who Are Changing Our World* (New York: Freeman, 2000), 196.

production of baked goods, paper, syrup, and juice   Felipe Sarmiento, Rocio Peralta, and Jenny M. Blamey, "Cold and Hot Extremozymes: Industrial Relevance and Current Trends," *Frontiers in Bioengineering and Biotechnology*, October 20, 2015, accessed February 28, 2017, https://www.ncbi .nlm.nih.gov/pmc/articles/PMC4611823/pdf/fbioe-03-00148.pdf.

The federal government was so interested   Accessed June 28, 2016, http://jgi .doe.gov/.

New England Biolabs   Accessed June 28, 2016, http://www.neb.uk.com /Katalogue-B/1/.

various industrial uses   Sarmiento, Peralta, and Blamey, "Cold and Hot Extremozymes."

"Do it in the dirt"   Norman Pace, interview with the author, Chicago, August 6, 1999. Unless otherwise noted, subsequent quotes of Pace are from this or another interview.

"bucketful of phenotypes"   Norman Pace, phone interview with the author, July 19, 1999. Unless otherwise noted, subsequent quotes of Pace are from this or another interview.

"transformed microbial ecology"   Gary Olsen, phone interview with the author, July 29, 1999. Unless otherwise noted, subsequent quotes of Olsen are from this interview.

"a kind of a pilgrimage"   David Stahl, phone interview with the author, August 3, 1999. Unless otherwise noted, subsequent quotes of Stahl are from this interview.

thirty-eight new species of Archaea   Susan M. Barns, Charles F. Delwiche, Jeffrey D. Palmer, and Norman R. Pace, "Perspectives on Archaeal Diversity, Thermophily and Monophyly from Environmental rRNA Sequences," *Proceedings of the National Academy of Sciences of the United States of America* 93, no. 17 (August 20, 1996): 9188–93, accessed August 19, 2016, http://www.pnas.org/content/93/17/9188.full.pdf.

than there were species of Archaea   Phil Hugenholtz, interview with the author, ASM meeting, Boston, June 18, 2016.

at a NASA conference in Chicago   Fox, interview, June 12, 2015.

Yonath wrote later   "Ada E. Yonath," accessed June 28, 2016, https://www .nobelprize.org/nobel_prizes/chemistry/laureates/2009/yonath-facts .html.

in her 2009 Nobel acceptance speech   Ibid.

By understanding the ribosome   Fox, interview, June 12, 2015.

Early in World War I   Fred Kelly, *One Thing Leads to Another: The Growth of an Industry* (New York: Houghton Mifflin, 1936), 27.

"Early life was not chaste"   Antonio Lazcano, "Natural History, Microbes and Sequences: Should We Look Back Again to Sequences?," *PLoS ONE* 6, no. 8 (August 16, 2011), accessed July 6, 2016, http://journals.plos.org/plosone/article?id=10.1371/journal.pone.0021334.

"microbial life invented the Internet"   Nigel Goldenfeld, phone interview with the author, January 22, 2016. Unless otherwise noted, subsequent quotes of Goldenfeld are from this interview.

at the bottom of the North Sea   Steffen L. Jørgensen, Ingunn H. Thorseth, Rolf B. Pedersen, Tamara Baumberger, and Christa Schleper, "Quantitative and Phylogenetic Study of the Deep Sea Archaeal Group in Sediments of the Arctic Mid-Ocean Spreading Ridge," *Frontiers in Microbiology* 4 (October 4, 2013), accessed February 6, 2017, http://www.ncbi.nlm.nih.gov/pmc/articles/PMC3790079/.

with no known precedents   Anja Spang, Jimmy H. Saw, Steffen L. Jørgensen, Katarzyna Zaremba-Niedzwiedzka, Joran Martijn, Anders E. Lind, Roel van Eijk, Christa Schleper, Lionel Guy, and Thijs J. G. Ettema, "Complex Archaea That Bridge the Gap between Prokaryotes and Eukaryotes," *Nature* 521 (May 14, 2015): 173–79, accessed February 7, 2017, doi:10.1038/nature14447.

that is characteristic of eukaryotes   Ibid.

"outgrowth of Archaea"   Brett Baker, phone interview with the author, July 28, 2016. Unless otherwise noted, subsequent quotes of Baker are from this interview.

quite probably all over the world   Spang et al., "Complex Archaea That Bridge the Gap between Prokaryotes and Eukaryotes."

industrial potential   Carolyn Marshall, "Bioprospectors Mine for Gold," accessed February 7, 2017, http://www.forbes.com/2000/05/29/feat.html.

Steven Benner, phone interview with the author, October 5, 2015.

Quoted in Martin Childs, "Carl Woese: Scientist Whose Work Revealed the Third Domain of Life," accessed February 2017, http://www.independent.co.uk/news/obituaries/professor-carl-woese-scientist-whose-work-revealed-the-third-domain-of-life-8521514.html.

"most important paper in the history of microbiology"   Jack Gilbert, phone interview with the author, January 12, 2016. Unless otherwise noted, subsequent quotes of Gilbert are from this interview.

"no matter what anybody else thinks"   Woese, quoted in Anton, *Bold Science*, 166.

Chapter Five

into the surrounding desert   "Shire of Murchison," Murchison Office, accessed July 1, 2016, http://www.murchison.wa.gov.au/.

did the Australian Astronomical Society issue a bulletin   "Meteoritical Bulletin: Entry for Murchison," *Meteoritical Society*, accessed July 1, 2016, http://www.lpi.usra.edu/meteor/metbull.php?code=16875.

researchers scrambled into action   Center for Meteorite Collection, accessed August 10, 2015, www.asu.edu.

city public high school   Sandra Pizzarello, interview with the author, Chicago, June 13, 2015. Unless otherwise noted, subsequent quotes of Pizzarello are from this or another interview.

the paper ignited controversy   John R. Cronin and S. Pizzarello, "Enantiomeric Excesses in Meteoritic Amino Acids," *Science* 275, no. 5302 (1997): 951–55, accessed July 14, 2016, bibcode:1997Sci . . . 275..951C, doi:10.1126/science.275.5302.951, PMID 9020072.

"a compelling hint"   David Levy, *Comets: Creators and Destroyers* (New York: Touchstone Books, 1998), 133.

Oxford's Balliol College   Robert Hazen, *The Story of Earth: The First 4.5 Billion Years, from Stardust to Living Planet* (New York: Random House, 2009), 69.

Sandra Pizzarello published   Sandra Pizzarello and Yongsong Huang, "The Deuterium Enrichment of Individual Amino Acids in Carbonaceous Meteorites: A Case for the Presolar Distribution of Biomolecule Precursors," *Geochimica et Cosmichimica Acta* 69, no. 3 (February 1, 2005): 599–605.

also subjected the rocks to hydrothermal pressure   Sandra Pizzarello, phone interview with the author, January 10, 2015. Unless otherwise noted, subsequent quotes of Pizzarello are from this or another interview.

"where you don't get . . . on site"   "Karen Meech," Solar System Exploration: Biographies, National Aeronautics and Space Administration, accessed July 1, 2016, https://solarsystem.nasa.gov/deepimpact/mission/bio -kmeech.cfm.

principal investigator, Don Brownlee   "Stardust: A Mission with Many Scientific Surprises," October 29, 2009, NASA Jet Propulsion Laboratory, accessed July 1, 2016, http://stardust.jpl.nasa.gov/news/news116.html.

vertical and even overhanging cliffs   Ibid.

Alcanivorax borkumensis   Samantha Joye, plenary speaker, "A Sea of Change: Altered Microbial Dynamics in the Wake of the Macondo Blowout," American Society for Microbiology General Meeting, New Orleans, May 31–June 2, 2015.

much more oil was spilling   Ibid.

the U.S. government claimed   John D. Sutter, "Defender of the Deep: The Oil's Not Gone," August 24, 2010, accessed June 27, 2016, http://www.cnn .com/2010/US/08/24/samantha.joye.gulf.oil/index.html; and Richard Harris, "Scientists Find Thick Layer of Oil on Seafloor," accessed February 9, 2017, http://www.npr.org/templates/story/story.php?storyId=129 782098&ft=1&f=1007.

"sudden microbial feeding frenzy" Joye, "A Sea of Change."

than previously thought   Nina Dombrowski, John A. Donaho, Tony Gutierrez, Kiley W. Seitz, Andreas P. Teske, and Brett J. Baker, "Reconstructing Metabolic Pathways of Hydrocarbon-Degrading Bacteria from the Deepwater Horizon Oil Spill," *Nature Microbiology* 1 (May 9, 2016), accessed February 9, 2017, https://www.researchgate.net/publication/302489116 _Reconstructing_metabolic_pathways_of_hydrocarbon-degrading _bacteria_from_the_Deepwater_Horizon_oil_spill.

"get so red"   Tullis Onstott, Skype interview with the author, April 7, 2016. Unless otherwise noted, subsequent quotes of Onstott are from this interview.

broadcast on the web and television   Web and TV live broadcast, www.philae .com or www.esa.com (European Space Agency), and on the Rosetta website, December 2014.

Kenneth and Juliette   Alexandre Bergatini and Ralf I. Kaiser, "In Situ Detection of Organics in the Comet 67P/Churyumov-Gerasimenko," *Chem* 1, no. 6 (December 8, 2016), accessed February 28, 2016, http://www .sciencedirect.com/science/article/pii/S2451929416302273.

"building blocks in space"   Michael Mumma, interview with the author, Astrobiology Science Conference, Chicago, June 13, 2015. Unless otherwise noted, subsequent quotes of Mumma are from this interview.

"here, you finish it"   Pizzarello, interview, June 10, 2015.

Chapter Six

coveted chemicals had come alive   Itay Budin, Skype interview with the author, February 3, 2016. Unless otherwise noted, subsequent quotes of Budin are from this interview.

helped shape the molecular-biological revolution   Alvin Powell, "Telomerase Work wins Szostak Nobel Prize in Medicine," *Harvard Gazette*, October 5, 2009, accessed July 5, 2016, http://news.harvard.edu/gazette/story /2009/10/jack-w-szostak-wins-nobel/.

to star death and planet birth   David Deamer, *First Life: Discovering the Con-*

nections Between Stars, Cells, and How Life Began (Berkeley and Los Angeles: University of California Press, 2011), 3.

more efficient versions of itself   Irene Chen, phone interview with the author, March 19, 2016. Unless otherwise noted, subsequent quotes of Chen are from this interview.

"by an ugly fact"   Friedrich Wöhler, quoted in "Urea and the Beginnings of Organic Chemistry," The Human Touch of Chemistry, accessed August 13, 2016, http://humantouchofchemistry.com/urea-and-the-beginnings-of -organic-chemistry.htm.

"fermentation and putrification"   M. C. Potter, "Electrical Effects Accompanying the Decomposition of Organic Compounds," *Proceedings of the Royal Society B: Biological Sciences* 84, no. 571 (September 14, 1911), accessed June 13, 2016, doi: 10.1098/rspb.1911.0073.

let him play with dangerous chemicals   Jack W. Szostak, "Jack W. Szostak: Biographical," Nobelprize.org, accessed July 5, 2016, http://www.nobelprize .org/nobel_prizes/medicine/laureates/2009/szostak-bio.html.

American national goal   Charles C. Price, "The New Era in Science," *Chemical and Engineering News* 43, no. 39 (1965): 90.

Harvard Medical School   "Research Spotlight: Jack Szostak," Origins of Life Initiative, Harvard University, accessed July 5, 2016, http://origins .harvard.edu/pages/research-spotlight-jack-szostak.

"the chemical building blocks of life"   J. Szostak, D. Bartel, and P. L. Luisi, "Synthesizing Life," *Nature* 409 (January 18, 2001): 387–90, doi:10.1038/ 35053176.

simpler than modern cells   I. A. Chen, R. W. Roberts, and J. W. Szostak, "The Emergence of Competition between Model Protocells," *Science* 305, no. 5689 (September 2004): 1474–77.

Luisi flew from Bangkok to Paro, where he was met   Pier Luigi Luisi, "The Origins of Life at a Buddhist Monastery in Bhutan," *Rendiconti Lincei: Scienzi fisiche e naturali* 24, no. 4 (December 2013): 387–400, accessed June 5, 2016, http://link.springer.com/article/10.1007/s12210-013-0259-8# /page-1.

"replace the entire petrochemical industry"   Craig Venter, "Craig Venter: On the Verge of Creating Synthetic Life," Ted Talks, YouTube.com, accessed July 5, 2016, https://www.youtube.com/watch?v=nKZ-GjSaqgo.

a job as director of communications   Anton, *Bold Science*, 30.

ballast to his driving personality   Jane Gitschier, "A Half-Century of Inspiration: An Interview with Hamilton Smith," *PLoS Genetics* 8, no. 1 (January 2012), accessed July 5, 2016, http://www.ncbi.nlm.nih.gov/pmc/articles /PMC3257296/pdf/pgen.1002466.pdf.

led two ocean expeditions   "The J. Robert Beyster and Life Technologies Foundation 2009–2010 Research Voyage of the Sorcerer II Expedition," J. Craig Venter Institute, accessed July 5, 2016, http://www.jcvi.org/cms /research/projects/gos/overview.

code to actively writing it   Venter, "Craig Venter: On the Verge of Creating Synthetic Life."

Barnet Cohen   Ibid.

Bruce Logan   Bruce E. Logan, "Research — Bioenergy: Microbial Fuel Cells," accessed July 5, 2016, Penn State, www.engr.psu.edu/ce/enve/logan /bioenergy/research_mfc.htm.

B. H. Kim   Ibid.

hundreds of thousands of failed attempts   Craig Venter, *Life at the Speed of Light: From the Double Helix to the Dawn of Digital Life* (New York: Viking, 2013), 204.

toxic waste cleanup   Wil S. Hylton, "Craig Venter's Bugs Might Save the World," *New York Times Magazine*, May 30, 2012, accessed July 5, 2016, http://www.nytimes.com/2012/06/03/magazine/craig-venters-bugs -might-save-the-world.html?_r=0.

"energy production"   H. O. Smith, R. M. Friedman, and J. C. Venter, "Biological Solutions to Renewable Energy," *Bridge* 33, no. 2 (July 1, 2003): 36–40, accessed March 20, 2016, http://www.jcvi.org/cms/publications /listing/abstract/article/biological-solutions-to-renewable-energy/.

Synbio (for "synthetic biology")   "The Activity Hub for the Synthetic Biology Industry," Synbiobeta, accessed February 16, 2017, http://synbiobeta .com/.

Asian Pacific Basin   "Synthetic Biology Market Increasing Research and Development Activities Lead to Remarkable Growth," Transparency Market Research, August 25, 2016, accessed August 29, 2016, www .openpr.com.

Abbvie   Luke Timmerman, "Abbvie, Maker of World's No. 1 Drug, Bets on Synthetic Biology Start-up," February 10, 2016, accessed August 25, 2016, http://www.forbes.com/sites/luketimmerman/2016/02/10/abbvie -maker-of-worlds-no-1-drug-bets-synthetic-biology-startup-can-keep -it-going/#4cdf383b72df.

"world of difference"   Jack Szostak, "Out of a Harsh and Hostile Earth," *New Scientist*, August 7, 2010, 26; published online as "Recreate Life to Understand How Life Began," accessed February 10, 2017, https://www .newscientist.com/article/mg20727721.100-recreate-life-to-understand -how-life-began/.

NASA's Steven Benner   Steve Benner, phone interview with the author, October 5, 2015. Unless otherwise noted, subsequent quotes of Benner are from this or another interview.

incredibly optimistic   Sara Walker, quoted in Cara Santa Maria, "Life Redefined: Is Information Processing the Key to Understanding How Life Will Arise Elsewhere in The Cosmos?," *Huffington Post*, December 24, 2012, accessed February 10, 2017, http://www.huffingtonpost.com/2012 /12/24/life-redefined-information-processing_n_2357987.html.

"it's the Wild West out there"   Mark Lyte, phone interview with the author, July 15, 2015. Unless otherwise noted, subsequent quotes of Lyte are from this or another interview.

World Health Organization Director Margaret Chan   "WHO Multi-Country Survey Reveals Widespread Public Understanding about Antibiotic Resistance," World Health Organization, November 16, 2015, accessed July 5, 2016, http://www.who.int/mediacentre/news/releases/2015/antibiotic -resistance/en/.

"the biggest health challenge of the 21st century"   "AMA Continues Efforts to Combat Antibiotic Resistance," American Medical Association, November 16, 2015, accessed July 5, 2016, http://www.ama-assn.org/ama/pub /news/news/2015/2015-11-16-combat-antibiotic-resistance.page.

Laurie Cox   Laurie Cox, Skype interview with the author, March 15, 2016. Unless otherwise noted, subsequent quotes of Cox are from this or another interview.

babies with those of Malawians in Africa   Tanya Yatsunenko and Martin Blaser, keynote presentation, American Society for Microbiology General Meeting, New Orleans, May 31–June 2, 2015.

caused by prenatal infections   Melissa Wenner, "Infected with Insanity: Could Microbes Cause Mental Illness?" *Scientific American Mind*, April/May 2008, accessed February 10, 2017, https://www.scientificamerican.com /article/infected-with-insanity/.

experienced better emotional health   Javier A. Bravo, Paul Forsythe, Marianne V. Chew, Emily Escaravage, Hélène M. Savignac, Timothy G. Dinan, John Bienenstock, and John F. Cryan, "Ingestion of *Lactobacillus* Strain Regulates Emotional Behavior and Central GABA Receptor Expression in a Mouse via the Vagus Nerve," *Proceedings of the National Academy of Sciences of the United States of America* 108, no. 38 (September 20, 2011): 16050–55.

from $11 billion to an astounding $32 billion   Scalar Market Research, "Synthetic Biology Market Worth 37.21 Billion USD by 2022," accessed February 21, 2017, www.scalarmarketresearch.com.

"Maybe we need to start again"   Benner, interview, October 5, 2015.

Chapter Seven

his hometown of Wernersville, Pennsylvania   Andrew Knoll, phone interview with the author, June 9, 2014. Unless otherwise noted, subsequent quotes of Knoll are from this or another interview.

convinced that life and Earth coevolved   Ibid.

in backyard soils   A. M. Gounot, "Microbial Oxidation and Reduction of Manganese: Consequences in Groundwater and Applications," *FEMS Microbiology Reviews* 14, no. 4 (August 1994): 339–49, accessed August 23, 2016, http://www.ncbi.nlm.nih.gov/pubmed/7917421.

whooshing "pop" noise   Robert Hazen, phone interview with the author, August 7, 2014. Unless otherwise noted, subsequent quotes of Hazen are from this or another interview.

"Darwin's 'descent with modification'"   Andrew Knoll, phone interview with the author, October 19, 2016. Unless otherwise noted, subsequent quotes of Knoll are from this or another interview.

working with that rover   Ibid.

Hazen imitated   "Mineral in Mars 'Berries' Adds to Water Story," Mars Exploration Rovers, accessed October 8, 2016, http://mars.nasa.gov/mer/news room/pressreleases/20040318a.html.

"winnowing them down"   Hazen, interview, August 9, 2016.

during Hazen's first experiment   Hazen, *The Story of Earth*, 161–62.

Nick Lane suggested   Nick Lane, *The Vital Question: Energy, Evolution, and the Origins of Complex Life* (New York: Viking Press, 2015), 211.

Hazen responded with a gusto   Robert Hazen, interview with the author, October 20, 2014.

stability of amino acids   Hazen, interview, August 7, 2014.

all the Earth's oceans   Becky Oskin, "Rare Diamond Confirms That Earth's Mantle Holds an Ocean's Worth of Water," *Scientific American*, Live Science, March 12, 2014, accessed October 8, 2016, https://www.scientific american.com/article/rare-diamond-confirms-that-earths-mantle -holds-an-oceans-worth-of-water/.

a dozen companies and researchers   Surjit Das, ed., Microbial Biodegradation and Bioremediation (London: Elsevier, 2014).

recovery (11 percent) or to municipal landfills (54 percent)   "Advancing Sustainable Materials Management: Facts and Figures Report," U.S. Environmental Protection Agency, 2012, accessed July 7, 2016, https://www .epa.gov/smm/advancing-sustainable-materials-management-facts-and -figures-report.

"positive feedback with the biosphere"   Hazen, interview, August 7, 2014.

Chapter Eight

"wondrous life"   Andrew Marvell, "The Garden" (1681), accessed July 6, 2016, http://www.poetryfoundation.org/poems-and-poets/poems/detail/44682.

trillion billion useful microbes   "A Gram of Soil," accessed July 10, 2016, http://nematode.unl.edu/gramsoil.htm; and "Biology Life in Soil," Soil Science Society of America, accessed June 5, 2014, http://www.soils4teachers.org/biology-life-soil.

"as the one in our guts"   Mike Amaranthus and Bruce Allyn, "Healthy Soil Microbes, Healthy People," *Atlantic*, June 11, 2013, accessed March 13, 2017, https://www.theatlantic.com/health/archive/2013/06/healthy-soil-microbes-healthy-people/276710/.

new soil microbe industry   J. Jacob Parnell, Randy Berka, Hugh A. Young, Joseph M. Sturino, Yaowei Kang, D. M. Barnhart, and Matthew V. DiLeo, "From the Lab to the Farm: An Industrial Perspective of Plant Beneficial Microorganisms," *Frontiers in Plant Science*, August 4, 2016, accessed March 13, 2017, http://journal.frontiersin.org/article/10.3389/fpls.2016.01110/full.

"wormery"   "Darwin's 'Big Soil Idea,'" Earth Learning Idea, accessed July 10, 2016, http://www.earthlearningidea.com/PDF/58_Darwin_worms.pdf.

"in the history of the world"   Charles Darwin, *The Formation of Vegetable Mould Through the Action of Worms with Observations on their Habits* (New York: Bookworm, 1976), 305.

many of our medicines   Anna Azvolinsky, "New Antibiotic from Soil Bacteria," *Scientist*, January 7, 2015, accessed July 10, 2016, http://www.the-scientist.com/?articles.view/articleNo/41850/title/New-Antibiotic-from-Soil-Bacteria/.

created twenty-two different antibiotic compounds   "Medicine from the Soil: Actinomycetes," Fairfax County, VA, accessed July 10, 2016, http://www.fairfaxcounty.gov/nvswcd/newsletter/soilmedicine.htm.

focused on another obscure microbe   C. A. Lowry, J. H. Hollis, A. de Vries, B. Pan, L. R. Brunet, J. R. F. Hunt, J. F. R. Paton, E. van Kampen, D. M. Knight, A. K. Evans, G. A. W. Rook, and S. L. Lightman, "Identification of an Immune-Responsive Mesolimbocortical Serotonergic System: Potential Role in Regulation of Emotional Behavior," *Neuroscience* 146, nos. 2–5 (May 11, 2007): 756–72, accessed February 22, 2016, doi:10.1016/j.neuroscience.2007.01.067.

"area no one was investigating"   Susan Jenks, phone interview with the

author, May 23, 2015. Unless otherwise noted, subsequent quotes of Jenks are from this interview.

"seemed to misunderstand that" Susan M. Jenks and Dorothy Matthews, "Ingestion of *Mycobacterium vaccae* Influences Learning and Anxiety in Mice," paper presented at the Annual Animal Behavior Society Meeting, William and Mary College, Williamsburg, VA, July 25–30, 2010.

Gardening Know How Bonnie L. Grant, "Antidepressant Microbes in Soil: How Dirt Makes You Happy," Gardening Know How, accessed July 10, 2016, http://www.gardeningknowhow.com/garden-how-to/soil-fertil izers/antidepressant-microbes-soil.htm.

"Is Dirt the New Prozac?" Josie Glausiusz, "Is Dirt the New Prozac?," *Discover Magazine*, June 14, 2007, accessed July 10, 2016, http://discover magazine.com/2007/jul/raw-data-is-dirt-the-new-prozac.

Healing Landscapes website "It's In the Dirt! Bacteria in Soil May Make Us Happier, Smarter," Therapeutic Landscapes Network, accessed July 10, 2016, http://www.healinglandscapes.org/blog/2011/01/its-in-the-dirt -bacteria-in-soil-makes-us-happier-smarter/.

lessened anxiety D. M. Matthews and S. M Jenks, "Ingestion of Mycobac-terium vaccae Decreases Anxiety-Related Behavior and Improves Learn-ing in Mice," *Behavioral Processes* 96 (June 2013): 27–35, accessed Febru-ary 23, 2017, https://www.ncbi.nlm.nih.gov/pubmed/23454729.

had improved cognitive function Emeran A. Mayer, Rob Knight, Sarkis K. Mazmanian, John F. Cryan, and Kirsten Tillisch, "Gut Microbes and the Brain: Paradigm Shift in Neuroscience," *Journal of Neuroscience* 34, no. 46 (November 12, 2014): 15490–96, accessed July 10, 2016, doi: 10.1523/ JNEUROSCI.3299-14.2014.

small working-class suburb Mark Lyte, phone interview with the author, July 16, 2015. Unless otherwise noted, subsequent quotes of Lyte are from this or another interview.

made for a real-life experiment James T. Rosenbaum, "The E. Coli Made Me Do It," *New Yorker*, November 8, 2013, accessed July 10, 2016, http:// www.newyorker.com/tech/elements/the-e-coli-made-me-do-it.

Lyte's general observations Mark Lyte, phone interview with the author, August 12, 2015. Unless otherwise noted, subsequent quotes of Lyte are from this or another interview.

sought to understand why Peter Andrey Smith, "Can the Bacteria in Your Gut Explain Your Mood?," *New York Times Magazine*, June 23, 2015.

in a national park Yong, *I Contain Multitudes*, 65.

typical Los Angeles high school student Elaine Hsiao, phone interview with

the author, July 9, 2015. Unless otherwise noted, subsequent quotes of
Hsiao are from this interview.

two phyla   Jennifer G. Mulle, William G. Sharp, and Joseph F. Cubells, "The
Gut Microbiome: A New Frontier in Autism Research," *Current Psychiatry
Reports* 15, no. 2 (February 2013): 337.

Emily Balskus   Emily Balskus, Skype interview with the author, Cambridge,
MA, July 18, 2015. Unless otherwise noted, subsequent quotes of Balskus
are from this or another interview.

immune system and emotional wellness   Vanessa Ridaura and Yasmine Belkaid, "Gut Microbiota: The Link to Your Second Brain," *Cell* 161, no. 2
(April 9, 2015): 193–94.

Israeli team   Ron Sender, Shai Fuchs, and Ron Milo, "Revised Estimates for
the Number of Human and Bacteria Cells in the Body," BioRxiv, posted
January 6, 2016, accessed July 5, 2016, http://biorxiv.org/content
/biorxiv/early/2016/01/06/036103.full.pdf.

"uncharted territory"   Weinstock, quoted in "Finally, a Map of All the
Microbes on Your Body," June 13, 2012, accessed July 5, 2016, http://
www.npr.org/sections/health-shots/2012/06/13/154913334/finally
-a-map-of-all-the-microbes-on-your-body.

"Digestive bliss"   Accessed August 15, 2016, https://www.activeadvantage
plus.com/product/probiotic-advantage-bifido-beadlet-50?key=208085.

"in every probiotics capsule"   Accessed August 14, 2016, https://www.amazon
.com/Probiotics-Billion-Capsule-Pro-Health-Naturenetics/dp/B00I3TFP
CI/?gclid=COP4z4D8wM4CFU5bhgodYNIOrw.

one or one-and-a-quarter to one   Ron Sender, Shai Fuchs, and Ron Milo,
"Revised Estimates for the Number of Human and Bacteria Cells in
the Body," *PLOS Biology* (August 19, 2016), accessed February 23, 2017,
https://doi.org/10.1101/036103, doi: 10.1371/journal.pbio.1002533.

for resistant infections, vancomycin   Rachel Feltman, "New Class of Antibiotic
Found in Dirt Could Prove Resistant to Resistance," *Washington Post*,
January 7, 2015, accessed July 10, 2016, https://www.washingtonpost
.com/news/speaking-of-science/wp/2015/01/07/new-class-of-anti
biotic-found-in-dirt-could-prove-resistant-to-resistance/.

"microbial biology without the microbiologist"   Slava Epstein, Skype interview
with the author, July 28, 2016. Unless otherwise noted, subsequent
quotes of Epstein are from this or another interview.

its taste and kick   Liz French, interview with the author, August 15, 2016.

"next round of evolution"   Jameson Rogers, quoted in Kat J. McAlpine, "How
New Biosensors turn *E. coli* into Something Valuable," *Harvard Gazette*,

August 4, 2015, accessed August 23, 2016, http://news.harvard.edu
/gazette/story/2015/08/how-new-biosensors-turn-e-coli-into-some
thing-valuable/; and McAlpine, "Bacteria 'Factories' Churn Out Valuable
Chemicals," *Harvard Gazette*, December 24, 2014, accessed October 12,
2016, http://news.harvard.edu/gazette/story/2014/12/bacteria-churn
-out-valuable-chemicals/.

Illumin   University of Southern California, accessed July 10, 2016, http://
illumin.usc.edu/about/.

Emefcy   Emefcy-Bio Energy Systems, accessed October 7, 2016, http://www
.emefcy.com/.

Pilus Energy   Pilus Energy LLC, accessed October 9, 2016, https://www
.linkedin.com/company/pilus-energy-llc.

called the EcoVolt   Demonstrateur EcoVolt, accessed October 10, 2016,
https://vimeo.com/68853154.

Widmer Brothers Brewing   Widmer Brothers Brewing, accessed October 10,
2016, http://widmerbrothers.com/history/.

could do so   "Scientists Discover Hazardous Waste-Eating Bacteria," September 14, 2014, accessed August 14, 2016, http://www.dalton.manchester
.ac.uk/news-and-events/scientists-discover-hazardous-waste-eating
-bacteria.htm; and John D. Coates, phone interview with the author,
February 21, 2017.

made beer from corn   Heather Whipps, "Beer Brewed Long Ago by Native
Americans," *Live Science*, December 28, 2007, accessed July 10, 2016,
http://www.livescience.com/4770-beer-brewed-long-native-americans
.html.

especially our beer   "Mourt's Relation: A Journal of the Pilgrims at Plymouth,
1622, Part I," The Plymouth Colony Archive Project (2007), accessed July
10, 2016, http://www.histarch.illinois.edu/plymouth/mourt1.html.

Chapter Nine

Michael Russell   Michael Russell, phone interview with the author, July 18,
2014. Unless otherwise noted, subsequent quotes of Russell are from this
or another interview.

Ireland's Silvermines deposits   Loreto Farrell, "Rehabilitation at Silvermine,"
Department of the Marine and Natural Resources, accessed July 13, 2016,
http://www.mineralsireland.ie/NR/rdonlyres/06497A37-028E-49E9
-A094-9DED962D286C/0/Rehabilitationatsilvermines.pdf.

Laurie Barge   "Laurie Barge," Curriculum vitae, Jet Propulsion Laboratory,

California Institute of Technology, accessed July 13, 2016, https://science
.jpl.nasa.gov/people/Barge/.

botanist Bill Martin   Bill Martin, "Early Evolution without a Tree of Life,"
Biology Direct 6, no. 36 (2011), accessed July 13, 2016, doi: 10.1186/1745
-6150-6-36.

popular videos and books   Nick Lane, *The Vital Question: Why Is Life the Way It
Is?* (London: Profile Books, 2015).

closest star   Kenneth Chang, "Another Earth Might Be Near, a Mere 25 Trillion Miles Away," *New York Times*, August 25, 2016, A11.

"not a scientist, never will be"   Michael Russell, Skype interview with the
author, August 10, 2015. Unless otherwise noted, subsequent quotes of
Russell are from this or another interview.

"These things are hollow!"   Michael Russell, phone interview with the author,
December 4, 2016. Unless otherwise noted, subsequent quotes of Russell
are from this or another interview.

plain of dolomite chimneys   John Whitfield, "Origin of Life: Nascence Man,"
Nature 459 (2009): 316.

in plants was a lever   Nick Lane, "Why Are Cells Powered by Proton Gradients?," *Nature Education* 3, no. 9 (2010): 18.

in the vent water   Whitfield, "Origin of Life."

John Sutherland   John Sutherland, phone interview with the author, May
12, 2015. Unless otherwise noted, subsequent quotes of Sutherland are
from this or another interview.

Published in Nature   Matthew W. Powner, Beatrice Garland, and John
Sutherland, "Synthesis of Activated Pyrimidine Ribonucleotides in Pre-
biotically Plausible Conditions," *Nature* 459 (May 14, 2009), 239–42.

Jupiter and Saturn   Wade, "Chemist Shows How RNA Can Be the Starting
Point."

all at one time   Bhavesh H. Patel, Claudia Percivalle, Dougal J. Ritson,
Colm D. Duffy, and John D. Sutherland, "Common Origins of RNA, Pro-
tein and Lipid Precursors in a Cyanosulfidic Proto-Metabolism," *Nature
Chemistry* 7 (2015): 301–7.

"like Middle Earth"   Quoted in Charles Q. Choi, "Life's Building Blocks Form
in Replicated Deep Sea Vents," *Astrobiology Magazine*, March 7, 2016,
accessed May 1, 2016, http://www.astrobio.net/origin-and-evolution-of
-life/lifes-building-blocks-form-in-replicated-deep-sea-vents/.

mind-blowing, dangerous thought!   Sutherland, interview, May 12, 2015.

"not a biologist"   Bill Martin, phone interview with the author, August 22,
2016. Unless otherwise noted, subsequent quotes of Martin are from this
or another interview.

other places in the solar system   Laurie Barge, phone interview with the author, December 15, 2016.

arsenic in place of phosophorus   Stolz, interview, December 9, 2015.

Twenty Questions   Whitfield, "Origin of Life," 319.

"a place of major revelation"   Bill Martin, phone interview with the author, December 20, 2016. Unless otherwise noted, subsequent quotes of Martin are from this or another interview.

"most of our computer server"   Bill Martin, phone interview with the author, August 10, 2016. Unless otherwise noted, subsequent quotes of Martin are from this or another interview.

"Our model is much simpler"   Martin, interview, December 20, 2016.

"won't be tidy"   Sutherland, interview, October 28, 2016.

Chapter Ten

whether her project would work   Margaret McFall-Ngai, Skype interview with the author, November 22, 2015. Unless otherwise noted, subsequent quotes of McFall-Ngai are from this or another interview.

at sunset it awoke   "*Euprymna scolopes*, Hawaiian Bobtail Squid," The Cephalopod Page, accessed December 15, 2015, http://www.thecephalopodpage .org/Escolopes.php.

and loved decoding the rules   Margaret McFall-Ngai, interview with the author, Boston, June 17, 2016. Unless otherwise noted, subsequent quotes of McFall-Ngai are from this or another interview.

logical and compelling   Margaret McFall-Ngai, phone interview with the author, March 21, 2015. Unless otherwise noted, subsequent quotes of McFall-Ngai are from this or another interview.

the sound of authority   Jamie Foster, phone interview with the author, January 30, 2015. Unless otherwise noted, subsequent quotes of Foster are from this interview.

she gave a presentation   Ned Ruby, phone interview with the author, February 12, 2015.

was to study   McFall-Ngai, interview, March 21, 2015.

scoop the muck at Yellowstone   Pace, interview, July 19, 1999.

"associated microbes"   Margaret McFall-Ngai, Michael G. Hadfield, Thomas C. G. Bosch, Hannah V. Carey, Tomislav Domazet-Ložo, Angela E. Douglas, Nicole Dubilier, et al., "Animals in a Bacterial World: A New Imperative for the Life Sciences," *Proceedings of the National Academy of Sciences of the United States of America* 110, no. 9 (February 26, 2013): 3229–36.

"discussed in polite circles"   Margaret McFall-Ngai, phone interview with the author, March 12, 2015.

for the first time to recognize   McFall-Ngai et al., "Animals in a Bacterial World."

"every microbiologist she could find"   McFall-Ngai, interview, March 21, 2015.

idea opened a door   Ruby, interview, February 12, 2015.

fell in love   Ibid.

"symbiosis the two of them have"   Nicole Dubilier, quoted in Ed Yong, "Here's Looking at You, Squid," *Nature* 517 (January 15, 2015): 263–65.

never been seen before   Anton, *Bold Science*, 201.

Santa Monica beach   McFall-Ngai, interview, June 17, 2016.

she invited about two dozen researchers   Ibid.

Polly Matzinger was proposing   Anton, *Bold Science*, 93.

"standing the immune-system theory on its head"   Claudia Dreifus, "A Conversation with Polly Matzinger: Blazing an Unconventional Trail to a New Theory of Immunity," *New York Times*, June 16, 1998, accessed July 16, 2016, http://www.nytimes.com/1998/06/16/science/conversation-with-polly-matzinger-blazing-unconventional-trail-new-theory.html.

composite of microbial and human cells and genes   The Gordon Lab, accessed July 10, 2016, https://gordonlab.wustl.edu/.

established in 2011   McFall-Ngai, interview, June 17, 2016.

medicinal products   "Maring Microbiome Conference, 2016," Berlin, accessed February 24, 2017, www.macumbaproject.edu. Subsequent notes are from this conference.

sea mammals and fish   Nicole Minadeo, Shedd Aquarium, interview with the author, February 27, 2017.

"tuning fork"   Jennifer G. Mulle, William G. Sharp, and Joseph F. Cubells, "The Gut Microbiome: A New Frontier in Autism Research," *Current Psychiatry Reports* 15, no. 2 (February 2015): 337.

process alcoholic drinks   T. Nosova, K. Jokelainen, P. Kaihovaara, H. Jousimies-Somer, A. Siitonen, R. Heine, and M. Salaspuro, "Aldehyde Dehydrogenase Activity and Acetate Production by Aerobic Bacteria Representing the Normal Flora of Human Large Intestine," *Alcohol Alcohol* 31, no. 6 (November 1996): 555–64, accessed August 15, 2016, http://www.ncbi.nlm.nih.gov/pubmed/9010546.

shared a core microbiome   Margaret McFall-Ngai, "Are Biologists in 'Future Shock'? Symbiosis Integrates Biology across Domains," *Nature Reviews Microbiology* 6 (October 2008): 789–92, accessed June 10, 2016, http://www.nature.com/nrmicro/journal/v6/n10/full/nrmicro1982.html.

virulence of hospital biofilms   Karen Visick, phone interview with the author, April 21, 2015.

"new imperative for the life sciences"   McFall-Ngai et al., "Animals in a Bacterial World."

"something controversial or new"   McFall-Ngai, interview, March 21, 2015.

"going over to the dark side"   Ibid.

"nested ecosystems"   Ibid.

"around the surrounding farms"   Ibid.

"name me one that is not"   Ibid.

"She doesn't deal in little ideas"   Yong, "Here's Looking at You, Squid," 263.

David Sabatini   Accessed August 15, 2016, www.youtube.com.

"Dianne Newman called me"   Dianne Newman, phone interview with the author, September 24, 2014. Unless otherwise noted, subsequent quotes of Newman are from this or another interview with her.

Chapter Eleven

going out on a Saturday night   Doug Wood, phone interview with the author, September 14, 2015.

up from 12 percent to 30 percent   "US Obesity Levels, 1990–2014," ProCon .org, September 19, 2014, accessed July 19, 2016, http://obesity.procon .org/view.resource.php?resourceID=006026.

was doubling every generation   Center for Disease Control's Division of Diabetes Translation, "Long-Term Trends in Diabetes," April 2016, accessed July 19, 2016, http://www.cdc.gov/diabetes/data.

Home Microbiome Study   Home Microbiome Study, accessed December 5, 2016, http://homemicrobiome.com/.

were dying from them   "Antibiotic/Antimicrobial Resistance," Centers for Disease Control and Prevention, accessed July 19, 2016, https://www .cdc.gov/drugresistance/.

In New York, Peggy Lillis   "Peggy Lillis," Peggy Lillis Foundation, 2016, accessed July 19, 2016, http://peggyfoundation.org/story/peggy-lillis/.

In Texas, eight-year-old Nick Johnson   Marc Lallanilla, "New Bacteria Threaten Public Health," ABC News, November 9, 2004, accessed July 19, 2016, http://abcnews.go.com/Health/story?id=235781&page=1.

in India   Lei Chin, Randall Todd, Julia Kiehlbauch, Maroya Walters, and Alexander Kallen, "*Notes from the Field*: Pan-Resistant New Delhi Metallo-Beta-Lactamase-Producing *Klebsiella pneumonia*—Washoe County, Nevada, 2016." Morbidity and Mortality Weekly Report 66, no. 1 (Janu-

ary 13, 2017): 33, accessed March 2, 2017, http://dx.doi.org/10.15585
/mmwr.mm6601a7.

infected 15,000 Americans a year   "Nearly Half a Million Americans Suffered
from *Clostridium difficile* Infections in a Single Year," Centers for Disease
Control, February 15, 2015, accessed July 19, 2016, http://www.cdc.gov
/media/releases/2015/p0225-clostridium-difficile.html.

regarded the jar   Terrence Momaday, "Marshall's Hunch," *New Yorker*, September 20, 1993, accessed July 15, 2016, http://www.newyorker.com
/magazine/1993/09/20/marshalls-hunch.

"dead Helicobacter pylori"   David Y. Graham, "The Only Good *Helicobacter
pylori* Is a Dead *Helicobacter pylori*," *Lancet* 350, no. 9070 (July 5, 1997):
70–71.

"since well before we became humans"   Martin Blaser, *Missing Microbes: How
the Overuse of Antibiotics Is Fueling Our Modern Plagues* (New York: Pica-
dor, 2014), 13.

Silicon Valley–raised Californian   Cox, interview, March 15, 2016.

used short bursts of medicine   Ibid.

"now an endangered species"   Blaser, *The Missing Microbes*, 178.

Blaser told the FDA   Ibid., 225.

"new frontier"   Laurie Cox, phone interview with the author, November 2,
2016. Unless otherwise noted, quotes from Cox are from this or another
interview.

Novobiotic   Accessed August 18, 2016, http://www.novobiotic.com/.

"It's a black art"   Carolyn Brown, "Antibiotic Discovery Heralds New World
of Drugs," *Canadian Medical Association Journal* 187, no. 4 (February 2,
2015), accessed July 25, 2016, http://www.cmaj.ca/content/early/2015
/02/02/cmaj.109-4985.

"astounding potency"   Slava Epstein, Skype interview with the author, July
27, 2016. Unless otherwise noted, quotes of Epstein are from this or
another interview.

"a totally different universe"   Cox, interview, November 2, 2016.

a new tool called the isolation chip   Greg St. Martin, "Groundbreaking Northeastern Research Sweeps the Globe," News@Northeastern, January 9,
2015, accessed July 19, 2016, http://www.northeastern.edu/news/2015
/01/groundbreaking-northeastern-research-sweeps-the-globe/.

Kim Lewis told Nature   Heidi Ledford, "Promising Antibiotic Discovered in
Microbial 'Dark Matter,'" *Nature*, January 7, 2015, accessed July 19, 2016,
http://www.nature.com/news/promising-antibiotic-discovered-in
-microbial-dark-matter-1.16675.

range of other probiotics   "What Is Lactobacillus Acidophilus?," Healthline, http://www.healthline.com/health/what-is-lactobacillus-probiotic#2.

Evelo   Accessed July 20, 2016, http://evelobio.com/#evelo-biosciences.

Elpida   Accessed July 20, 2016, http://www.bio-elpida.com/en/facilities .htm.

Johnson and Johnson   Accessed July 20, 2016, http://www.jnj.com/.

"fast-moving field"   Cox, interview, March 15, 2016.

restore customers' libidos   Jonathan A. Eisen, "Today's Awful Overselling of the Microbiome: Robynne Chutkan on Libido," The Tree of Life, September 8, 2015, accessed August 21, 2016, https://phylogenomics.blogspot .com/2015/09/todays-awful-overselling-of-microbiome.html.

microbiome hype   Jonathan A. Eisen, "Today's Misleading Overselling the #Microbiome: U. Chicago on Alzheimer's and Gut Microbes," The Tree of Life, July 22, 2016, accessed August 18, 2016, https://phylogenomics .blogspot.com/2016/07/todays-misleading-overselling.html?utm _source=feedburner&utm_medium=feed&utm_campaign=Feed:+The TreeOfLife+(The+Tree+of+Life).

Jeffrey Gordon   The Gordon Lab, accessed September 1, 2016, https://gordon lab.wustl.edu/.

"a correction of the metabolic abnormalities"   Gina Kolata, "Gut Bacteria from Thin Humans Can Slim Mice Down," *New York Times*, September 5, 2013, accessed July 19, 2016, http://www.nytimes.com/2013/09/06/health /gut-bacteria-from-thin-humans-can-slim-mice-down.html.

"We are assimilated"   Actigenomics Team, "Prof. Jeffrey Gordon: 'The Gut Microbe May Drive a New Area of Precision Nutrition,'" October 15, 2012, accessed July 19, 2016, http://www.actigenomics.com/2012/10 /jeffrey-gordon/.

"synbiotics"   Michael Pollan, "Some of My Best Friends Are Germs," *New York Times Magazine*, May 15, 2013, accessed July 19, 2016, http://www .nytimes.com/2013/05/19/magazine/say-hello-to-the-100-trillion -bacteria-that-make-up-your-microbiome.html.

accelerate protein discovery   Accessed February 25, 2017, www.secondgenome .com.

Ana Swanson of the Washington Post   Ana Swanson, "What Tyson's Pledge to Stop Using Human Antibiotics in Chicken Means for the Future of Super-bugs," *Washington Post*, April 28, 2015, accessed July 19, 2016, https:// www.washingtonpost.com/news/wonk/wp/2015/04/28/what-tysons -pledge-to-stop-using-human-antibiotics-in-chicken-means-for-the -future-of-superbugs/.

Tyson   Ibid.

Jack Gilbert   "Jack A. Gilbert," Argonne National Laboratory, accessed July 19, 2016, http://www.anl.gov/contributors/jack-gilbert.

managed the Earth Microbiome Project   Jack Gilbert, interview with the author, University of Chicago Hospital, September 21, 2015.

environments around the world   "Jack Gilbert, Ph.D.," Ecology of Evolution, University of Chicago, accessed July 19, 2016, http://pondside.uchicago .edu/ecol-evol/people/gilbert.html.

Launched in 2010   Earth Microbiome Project, accessed December 6, 2016, http://www.earthmicrobiome.org/.

one of three ways   Antimicrobial Resistance Learning Site for Veterinarians, accessed June 27, 2016, http://amrls.cvm.msu.edu/microbiology /molecular-basis-for-antimicrobial-resistance/acquired-resistance /acquisition-of-antimicrobial-resistance-via-horizontal-gene-transfer.

one October day   Doug Wood, phone interview with the author, March 14, 2016. Unless otherwise noted, subsequent quotes of Wood are from this or another interview.

A month later Wood's son Ben   Doug Wood, interview with the author, September 14, 2015. Unless otherwise noted, subsequent quotes of Wood are from this or another interview.

Older people's gut microbiomes   Ibid.

Dianne K. Newman, at Caltech   Newman Lab, accessed August 14, 2016, http://dknweb.caltech.edu/Newman_Lab.html.

"a problem they have to solve"   Dianne Newman, phone interview with the author, September 25, 2014. Unless otherwise noted, subsequent quotes of Newman are from this or another interview with her.

Perdue had been perfecting   Stephanie Strom, "Perdue Sharply Cuts Antibiotic Use in Chickens and Jabs at Its Rivals," *New York Times*, July 31, 2015, accessed July 19, 2016, http://www.nytimes.com/2015/08/01/business /perdue-and-the-race-to-end-antibiotic-use-in-chickens.html?_r=0.

sponsored by the Obama White House   "Fact Sheet: Over 150 Animal and Health Stakeholders Join White House Effort to Combat Antibiotic Resistance," White House, June 2, 2015, accessed July 19, 2016, https://www .google.com/url?sa=t&rct=j&q=&esrc=s&source=web&cd=1&ved=0ahUK Ewj9zdr9gYDOAhVh_4MKHZYoAGEQFggeMAA&url=https%3A%2F%2F www.whitehouse.gov%2Fsites%2Fdefault%2Ffiles%2Fdocs%2F060215 _private_sector_factsheet.pdf&usg=AFQjCNGhEoR7IP1oXPQAtLwlpc30 KvDV-w&sig2=syuk2-6r6MKcB8P5LAy_WQ&cad=rjt.

Rodney Dietert   Rodney Dietert, *The Human Superorganism: How the Microbiome Is Revolutionizing the Pursuit of a Healthy Life* (New York: Dutton, 2016), 245.

Chapter Twelve

September 28, 2015 "Discovery of Water on Mars Boosts Possibility of Life on Planet," *Chicago Tribune*, September 28, 2015.

the region known as Utopia Planitia Dwayne Brown, Laurie Cantillo, Guy Webster, and Anton Caputo, "Mars Ice Deposit Holds as Much Water as Lake Superior," November 22, 2016, accessed December 1, 2016, https:// www.nasa.gov/feature/jpl/mars-ice-deposit-holds-as-much-water-as -lake-superior.

Facebook pages https://www.facebook.com/MarsCuriosity/videos/9237468 71008622/, https://www.facebook.com/NASA/.

Old Dominion University's Nora Noffke Nora Noffke, "Ancient Sedimentary Structures in the <3.7 Ga Gillespie Lake Member, Mars, That Resemble Macroscopic Morphology, Spatial Associations, and Temporal Succession in Terrestrial Microbialites," *Astrobiology* 15, no. 2 (February 2015): 169–202, doi: 10.1089/ast.2014.1218.

a sweltering room in Pasadena Gil Levin, phone interview with the author, October 13, 2015. Unless otherwise noted, subsequent quotes of Levin are from this interview.

Arguments intensified, however Steven Benner, phone interview with the author, November 10, 2014. Unless otherwise noted, subsequent quotes of Benner are from this or another interview.

testing positive for microbes Tamara Dietrich, "Did NASA Langley's Viking landers Find Life on Mars?," *Hampton (Virginia) Daily Press*, July 18, 2016, accessed July 18, 2016, http://www.dailypress.com/news/science/dp -nws-viking-life-20160717-story.html.

powerful oxidizing agents Chris McKay, phone interview with the author, July 19, 2016. Unless otherwise noted, quotes of McKay are from this interview.

"there is microbial life" Nicole Casal Moore, "Microbial Life on Mars: Could Saltwater Make It Possible?," *Michigan News*, August 16, 2011, accessed July 23, 2016, http://ns.umich.edu/new/releases/8510-microbial-life -on-mars-could-saltwater-make-it-possible.

University of Arizona undergraduate Ari Espinoza, "HiRISE Images Show Signs of Liquid Water on Mars," *UANews*, September 28, 2015, accessed July 23, 2016, https://uanews.arizona.edu/story/hirise-images-show -signs-of-liquid-water-on-mars.

NASA investigator Steven Benner Benner, interview, November 10, 2014.

Boston made the case for Mars Penelope Boston, Skype interview with the author, August 24, 2015. Unless otherwise noted, subsequent quotes of Boston are from this interview.

"very high level"   Penelope Boston, email to the author, June 29, 2016.

"but they're there"   Alison Murray, phone interview with the author, July 7, 2016. Unless otherwise noted, subsequent quotes of Murray are from this interview.

twenty-eight of which were newly discovered   Christopher T. Brown, Laura A. Hug, Brian C. Thomas, Itai Sharon, Cindy J. Castelle, Andrea Singh, Michael J. Wilkins, Kelly C. Wrighton, Kenneth H. Williams, and Jillian F. Banfield, "Unusual Biology across a Group Comprising More than 15% of Domain Bacteria," *Nature* 523 (July 9, 2015): 208–11.

hot coastal Virginia summer   Nora Noffke, phone interview with the author, February 6, 2015. Unless otherwise noted, subsequent quotes of Noffke are from this or another interview.

"I'll submit a hypothesis paper"   Ibid.

the Curiosity team reacted strongly   Nora Noffke, "Ancient Sedimentary Structures in the <3.7 Ga Gillespie Lake Member, Mars, That Resemble Macroscopic Morphology, Spatial Associations, and Temporal Succession in Terrestrial Microbialites," *Astrobiology* 15, no. 2 (February 2015): 169–92.

"clouds in the sky"   Linda Kah, Skype interview with the author, January 21, 2015. Unless otherwise noted, subsequent quotes of Kah are from this interview.

"to support life in the future"   Accessed March 13, 2017, https://www.nasa .gov/press-release/nasa-confirms-evidence-that-liquid-water-flows-on -today-s-mars.

"It's now an eroded hillside"   Noffke, phone interview, February 6, 2015.

only be produced   Greg Retallack, phone interview with the author, February 26, 2017.

after a rain   Gregory J. Retallack, "Paleosols and Paleoenvironments of Early Mars," *Geology* 42, no. 9 (September 2014): 755–58.

"But what exceeds all wonders"   Galileo, "Jupiter Quotes," accessed July 25, 2016, http://todayinsci.com/QuotationsCategories/J_Cat/Jupiter -Quotations.htm.

Europa featured sulfate salts   "A Window into Europa's Ocean Right at the Surface," NASA, March 5, 2013, accessed July 23, 2016, http://www.nasa .gov/topics/solarsystem/features/europa20130305.html.

younger member   Richard Greenberg, *Unmasking Europa: The Search for Life on Jupiter's Ocean Moon* (New York: Copernicus Books, 2008), 17.

through an Enceladus plume   "Deepest-Ever Dive through Enceladus Plume Completed," NASA, October 28, 2015, accessed July 10, 2016, http://www.jpl.nasa.gov/news/news.php?feature=4755.

methane-based life   McKay, interview, July 19, 2016.

Pluto's frigid ice surface   Bill Steigerwald, "Cracks in Pluto's Moon Could Indicate It Once Had an Underground Ocean," NASA, June 13, 2014, accessed July 23, 2016, http://www.nasa.gov/content/goddard/cracks-in-plutos-moon-could-indicate-it-once-had-an-underground-ocean.

more numerous than sunlike stars Michaël Gillon, Amaury H. M. J. Triaud, Brice-Olivier Demory, Emmanuël Jehin, Eric Agol, Katherine M. Deck, Susan M. Lederer, Julien de Wit, Artem Burdanov, James G. Ingalls, et al., "Seven Temperate Terrestrial Planets around the Nearby Ultracool Dwarf Star TRAPPIST-1," *Nature* 542 (February 23, 2017): 456–60.

better waste treatment   Coates, interview, February 21, 2017.

"boiling coffee in it!"   Irene Chen, email to the author, March 8, 2016.

Bayer, DuPont, and Monsanto   Sarah Zhang, "Good Riddance, Chemicals: Microbes Are Farming's Hot New Pesticides," *Wired*, March 21, 2016, accessed July 23, 2016, http://www.wired.com/2016/03/good-riddance-chemicals-microbes-farmings-hot-new-pesticides/.

plastic-degrading enzymes   Michelle Hampson, "*Science*: Newly Identified Bacteria Beat Down Tough Plastic," AAAS, March 9, 2016, accessed July 23, 2016, http://www.aaas.org/news/science-newly-identified-bacteria-break-down-tough-plastic.

"basic knowledge of life"   Ed Yong, "The Mysterious Thing about a Marvelous New Synthetic Cell," *Atlantic*, March 24, 2016, accessed July 23, 2016, http://www.theatlantic.com/science/archive/2016/03/the-quest-to-make-synthetic-cells-shows-how-little-we-know-about-life/475053/.

"an unprecedented step"   Quoted in Ed Yong, "The Mysterious Thing about a Marvelous New Synthetic Cell," *Atlantic*, March 24, 2016, accessed March 3, 2017, https://www.theatlantic.com/science/archive/2016/03/the-quest-to-make-synthetic-cells-shows-how-little-we-know-about-life/475053/.

"should be happy about"   Quoted in Erik Stokstad, "Engineered Crops Could Have It Made in the Shade," *Science* 354, no. 6314 (November 18, 2016): 816.

for such a long time   Freeman Dyson, "Green Universe," *New York Review of Books*, November 11, 2016, accessed November 29, 2016, http://www.nybooks.com/articles/2016/10/13/green-universe-a-vision/.

an international human microbiome effort   "A Biosphere without Borders,"

Max-Planck-Institut für Marine Mikrobiologie, October 28, 2015,
accessed July 24, 2016, http://www.mpi-bremen.de/en/A_biosphere
_without_borders.html.
"more news will follow"   Galileo, "Jupiter Quotes."

Chapter Thirteen

The hot sun blazed   Stolz, interview, September 28, 2016.
"almost exactly like Mono Lake water"   Ibid.
could neutralize radioactive waste   "Can Nuclear Waste Be Neutralized by Bacteria?," Engineering.com., September 12, 2014, accessed July 24, 2016,
http://www.engineering.com/DesignerEdge/DesignerEdgeArticles
/ArticleID/8471/Can-Nuclear-Waste-Be-Neutralized-by-Bacteria.aspx.
Heartland Microbes   , accessed July 22, 2016http://www.heartlandmicrobes
.com.
some of the most resistant pathogens   David Salisbury, "Life's Extremists May
Be an Untapped Source of Antibacterial Drugs," *Research News at Vanderbilt*, November 21, 2014, accessed April 23, 2015, https://news.vanderbilt
.edu/2014/11/21/lifes-extremists-may-be-an-untapped-source-of-anti
bacterial-drugs.
"veritable zoo"   Gaetan Bogonie, Skype interview with the author, April 10,
2016. Unless otherwise noted, subsequent quotes of Bogonie are from
this interview.
72,649 microbial genomes sequenced   National Center for Biotechnology
Information Database, accessed August 23, 2016, http://www.ncbi.nlm
.nih.gov/genome/browse/.
U.S. Naval research geologist Leonard Tender   Leonard M. Tender, "Microbial
Fuel Cells for Powering Navy Devices," Naval Research Laboratory, January 20, 2014, accessed July 21, 2016, www.ADA594746.pdf.
$38.7 billion a year by 2020   "Synthetic Biology Market Is Expected to Reach
$38.7 Billion, Globally, by 2020," *Bio-IT World*, June 25, 2014, accessed
July 24, 2016, http://www.bio-itworld.com/Press-Release/Synthetic
-Biology-Market-is-Expected-to-Reach-$38-7-Billion,-Globally,-by-2020
---Allied-Market-Research.
"rewriting agricultural history"   Steve Jordan, "Ag Giants Will Team Up in
'Game-Changing' Alliance," *Omaha World-Herald*, December 11, 2013,
accessed July 24, 2016, http://www.omaha.com/money/ag-giants-will
-team-up-in-game-changing-alliance/article_c1e5f5fb-649c-54df-adac
-00caacd59ef0.html.

"in my hand were converging"   John Stolz, Skype interview with the author, September 28, 2016. Unless otherwise noted, subsequent quotes of Stolz are from this or another interview.

"use arsenate and chlorate to grow"   Wenjie Sun, Reyes Sierra-Alvarez, Lily Milner, and Jim A. Field, "Anaerobic Oxidation of Arsenite Linked to Chlorate Reduction," *Applied and Environmental Microbiology* 76, no. 20 (October 2010): 6804–11, doi: 10.1128/AEM.00734-10.

"chance of microbe fossils"   Noffke, interview with the author, February 5, 2016.

Epcot Crenshaw   Epcot Crenshaw, accessed July 24, 2016, http://www.epcot crenshaw.com/.

"generating their own wastewater energy!"   Stolz, interview, September 28, 2016.

Emefcy   Emefcy-Bio Energy Systems, accessed November 11, 2016, http://www.emefcy.com/.

EcoVolt   Demonstrateur EcoVolt, accessed November 12, 2016, https://vimeo.com/68853154.

"similar to the first primordial mixture"   Heartland Microbes, accessed July 22, 2016, http://www.heartlandmicrobes.com.

Microbe Inotech   Microbe Inotech Labs, accessed November 9, 2016, http://www.microbeinotech.com/.

Service-Tech in Cleveland   Service-Tech Corporation, accessed February 11, 2017, http://www.service-techcorp.com/index.html.

(GEMC) in Delhi   "About iGEM," accessed November 7, 2016, http://igem.org/About.

"totally different microbial ecologies"   Slava Epstein, Skype interview with the author, July 26, 2016. Unless otherwise noted, quotes of Epstein are from this or another interview.

labs like that of Irene Chen   Chen, email to the author.

"Year of the Phage"   Accessed March 13, 2017, http://2015phage.org.

"it's not very hard to do"   Forest Rowher, phone interview with the author, April 27, 2016.

synthetic-biology team, headed by Pamela Silver   Wyss Institute for Biologically Inspired Engineering, accessed August 30, 2016, http://wyss.harvard.edu/viewpage/about-us/about;jsessionid=93362C6DA4 87906296A4C4740A559E15.wyss1.

extending healthful human life spans   "Healthcare, Tech and Longevity: Big Changes Ahead?," in-house science panel, MetLife, New York, August 19, 2016.

Unified Microbiome Initiative Consortium (UMIC)   Kat McAlpine, "Microbiomes Could Hold Keys to Improving Life," *Harvard Gazette*, October 30, 2015, accessed July 24, 2016, http://news.harvard.edu/gazette/story /2015/10/microbiomes-could-hold-keys-to-improving-life/.

University of Delaware researcher Jennifer Biddle   Curriculum vitae, accessed July 19, 2016, http://www1.udel.edu/experts/cvs/17786985385-Jennifer _Frances_Biddle.pdf.

"throw microbes under the bus"   Jennifer Biddle, phone interview with the author, July 26, 2016.

of the tropical rainforest   John Stolz, "Gaia and Climate Change," paper presented at the First American Geophysical Union Chapman Conference on the Gaia Hypothesis, San Diego, CA, March 7, 1988.

researchers at Miami University   Maribeth O. Hassett, Mark W. F. Fischer, and Nicholas P. Money, "Mushrooms as Rainmakers: How Spores Act as Nuclei for Raindrops," *PLOS One* 10, no. 10 (October 28, 2015), e0140407, accessed July 24, 2016, http://journals.plos.org/plosone/article?id=10 .1371/journal.pone.0140407#abstract0, doi: 10.1371/journal.pone .104047.

to sound a policy call   Ann Reid and Shannon E. Greene, "How Microbes Can Help Feed the World," report on an American Academy of Microbiology Colloquium, Washington, DC, December 2012, accessed July 24, 2016, http://academy.asm.org/index.php/browse-all-reports/800-how -microbes-can-help-feed-the-world.

Roser Matamala was the principal investigator   Accessed February 11, 2017, https://www.anl.gov/contributors/roser-matamala.

Some microbes could devour plastic   Uwe T. Bornscheuer, "Feeding on Plastic," *Science* 351, no. 6278 (March 11, 2016): 1154–55.

"We pulled from thirteen disciplines"   Dianne Newman, interview with the author, December 14, 2016. Unless otherwise noted, all quotes of Newman are from this or another interview.

rose by 485 percent   Andrew J. Ritter, "A Race to Turn Trillions of Our Own Bacteria into Medical Breakthroughs," CNBC, accessed November 29, 2016, http://www.cnbc.com/2016/11/29/microbiome-breakthroughs -creating-new-approach-to-medical-mysteries.html.

"future of the fragrance industry"   Aviva Rutkin, "Would You Wear Yeast Perfume? Microbes Used to Brew Scent," *New Scientist* 3011 (March 7, 2015), accessed December 2, 2016, https://www.newscientist.com/article/mg 22530113-600-would-you-wear-yeast-perfume-microbes-used-to-brew -scent/.

The new companies included   Accessed February 11, 2017, http://www .enterome.fr; accessed March 3, 2017, http://www.evelobio.com.

eight to ten genetically modified microbes   Antonio Regalado, "Companies Bet on Designer Bacteria as New Way to Treat Disease," *MIT Technology Review*, November 8, 2016, accessed December 1, 2016, https://www .technologyreview.com/s/602724/companies-bet-on-designer-bacteria -as-new-way-to-treat-disease/.

SER 109   Seres Therapeutics, accessed July 24, 2016, http://www.seres therapeutics.com/pipeline/ser-109.

Jack Gilbert's band Turnaround Time   Jack Gilbert, interview with the author, January 11, 2016.

"somebody is going to send me to Antarctica!"   Ibid.

comprise up to 20 percent of marine fuel   "Shipping and Biofuels: A Match Made in Heaven?," *Hellenic Shipping News Worldwide*, May 11, 2015, accessed July 24, 2016, http://www.hellenicshippingnews.com/shipping -and-biofuels-a-match-made-in-heaven/.

In Washington State, Hardwood Fuels   "Biofuel Projects," Bioenergy Washington, accessed October 7, 2016, http://www.bioenergy.wa.gov/biofuel projects.aspx.

Chapter Fourteen

"and it's beautiful"   Eric Boyd, personal conversation with the author, Yellowstone National Park, August 17, 2015.

"Jesus Christ!"   Melody Lindsay, personal conversation with the author, Bozeman, MT, August 17, 2015.

"Bring your bear spray"   Boyd, personal conversation, August 17, 2015.

sufferers from asthma   "Increasing Rates of Allergies and Asthma," American Academy of Allergy, Asthma, and Immunology, 2016, accessed August 4, 2016, https://www.aaaai.org/conditions-and-treatments/library/allergy -library/prevalence-of-allergies-and-asthma.

young women suffering from depression   Chris Iliades, "Stats and Facts about Depression in America," Everyday Health, 2013, accessed August 4, 2016, http://www.everydayhealth.com/hs/major-depression/depression -statistics/.

"100-day window"   "Four Gut Bacteria Decrease Asthma Risk in Infants," UBC News, September 30, 2015, accessed August 4, 2016, http://news .ubc.ca/2015/09/30/four-gut-bacteria-decrease-asthma-risk-in-infants/.

a wonder bug   Michael Pollan, *Cooked* [Netflix documentary], accessed May 20, 2016, https://www.netflix.com/title/80022456.

autism in children   Melinda Wenner Moyer, "Gut Bacteria May Play a Role in Autism," *Scientific American*, September 1, 2014, accessed August 4, 2016,

http://www.scientificamerican.com/article/gut-bacteria-may-play-a-role-in-autism/.

on the autism spectrum   Jack A. Gilbert, Rosa Krajmalnik-Brown, Dorota L. Porazinska, Sophie J. Weiss, and Rob Knight, "Toward Effective Probiotics for Autism and Other Neurodevelopmental Disorders," *Cell* 155, no. 7 (December 19, 2013): 1446–48, accessed August 2, 2016, http://www.cell .com/cell/pdf/S0092-8674(13)01486-4.pdf, doi: 10.1016/j.cell.2013.11 .035.

now looked more questionable   E. G. Severance, R. H. Yolken, and W. W. Eaton, "Autoimmune Diseases, Gastrointestinal Disorders and the Microbiome in Schizophrenia: More than a Gut Feeling," *Schizophrenia Research* 176, no. 1 (September 2016): 23–35.

"no healthy microbiome"   "Is a Messed-Up Microbiome Linked to Obesity? New U-M Study Casts Doubt," Michigan Medicine, University of Michigan, accessed August 24, 2016, http://www.eurekalert.org/pub_releases /2016-08/uomh-iam081916.php.

"We need a whole new language"   Jeff Miller, quoted in Alan Brown, "Microbial Manifesto: The Global Push to Understand the Microbiome (Kavli Roundtable), Live Science, March 2, 2016, accessed March 3, 2016, http:// www.livescience.com/53916-understanding-life-means-understanding -the-microbiome/.

"the key to ancient life"   Boyd, personal conversation, August 17, 2015.

growing up in Iowa   Ibid.

most important genes   Biddle, interview, July 26, 2016.

National Microbiome Initiative   "FACT SHEET: Announcing the National Microbiome Initiative," White House, May 13, 2016, accessed August 4, 2016, https://www.whitehouse.gov/the-press-office/2016/05/12/fact -sheet-announcing-national-microbiome-initiative.

wandered the gleaming hallways   Keynote address, ASM Microbe, annual meeting of the American Society for Microbiology, Boston, June 16–20, 2016.

twenty years ago   Carl Zimmer, "Triumph of the Archaea," *Discover*, February 1, 1995, accessed August 19, 2016, http://discovermagazine. com/1995/feb/triumphofthearch475.

"a new paradigm for understanding life on Earth"   Virginia Morell, "Microbial Biology: Microbiology's Scarred Revolutionary," *Science* 276, no. 5313 (May 2, 1997): 699–702.

"Under the rocks are the words"   Norman Maclean, *A River Runs Through It* (Chicago: University of Chicago Press, 1976), 233.

best diet for the person   "The Personalized Nutrition Project," American Committee for the Weizmann Institute of Science, accessed August 3, 2016, http://www.weizmann-usa.org/e-news/13-09/landing/personalized -nutrition-project.htm.

DNA for manmade cells "A New Assembly Line in London Is Making DNA," Futurism, accessed August 4, 2016, http://futurism.com/new-assembly -line-london-making-dna/.

researchers reversed photosynthesis "Biotech Breakthrough: Sunlight Can Be Used to Produce Chemicals and Energy," April 4, 2016, accessed August 2, 2016, http://www.science.ku.dk/english/press/news/2016/biotech -breakthrough-sunlight-can-be-used-to-produce-chemicals-and-energy /?ref=sfp-UK_biotech.

had not received microbiome transplants Brown, "Microbial Manifesto."

microbiologist James Liao Kyle Grice, phone interview with the author, August 1, 2016.

RNA had been shown to replicate itself Jesse Emspak, "'RNA World': Scientists Inch Closer to Recreating Primordial Life," August 19, 2016, accessed September 10, 2016, www.livescience.com.

Irene Chen Chen, email to the author.

Laurie Cox Cox, interview, November 2, 2016.

repair cracks in concrete "Self-Healing of Concrete by Bacterial Mineral Precipitation," accessed August 3, 2016, Delft University of Technology, March 2016, http://www.citg.tudelft.nl/en/research/projects/self -healing-concrete/.

tidewater soil Jim Shamp, "From Gene Editing to Microbiome, Precision Medicine to Big Data and More: NC's Bright Biotech Future," March 3, 2016, accessed June 20, 2016, http://wraltechwire.com/from-gene -editing-to-microbiome-precision.

"Harvard is investing in it" Gilbert, interview, February 10, 2017.

where researcher Aron Stubbins Mike Sullivan, "Scientists Research Deep-Sea Hydrothermal Vents, Find Carbon-Removing Properties," UGA Today, October 29, 2015, accessed August 4, 2016, http://news.uga.edu/releases /article/deep-sea-hydrothermal-vents-carbon-removing-properties -1015/.

"complex communities of beneficial microbes" Margaret McFall-Ngai, "Concept Adaptive Immunity: Care for the Community," *Nature* 445 (January 11, 2007): 153, accessed March 13, 2017, doi:10.1038/445153a.

"vastly transformative" Newman, interview, December 14, 2016.

Epilogue

Gulliver   Epstein, interview, July 26, 2016.

The new gene-editing tool   Yong, *I Contain Multitudes*, 261.

"skyscrapers everywhere"   Biddle, interview, July 25, 2016.

to lag behind the rates of support of some European and Asian governments
Sara Reardon, Jeff Tolleson, Alexandra Witze, and Erin Ross, "US Science Agencies Face Deep Cuts in Trump Budget," *Nature*, March 16, 2017, accessed March 30, 2017, http://www.nature.com/news/us-science -agencies-face-deep-cuts-in-trump-budget-1.21652.

"our geology and environment"   Woese, interview, July 14, 1999.

Index